Mammals

Mounta

George Olin

Alpha Editions

This edition published in 2022

ISBN : 9789356715745

Design and Setting By
Alpha Editions
www.alphaedis.com
Email - info@alphaedis.com

As per information held with us this book is in Public Domain. This book is a reproduction of an important historical work. Alpha Editions uses the best technology to reproduce historical work in the same manner it was first published to preserve its original nature. Any marks or number seen are left intentionally to preserve its true form.

ACKNOWLEDGEMENTS

With this booklet, as with *Mammals of the Southwest Deserts*, we are indebted to Dr. E. L. Cockrum, Assistant Professor of Zoology at the University of Arizona who has checked the manuscript for accuracy. We are also grateful to him for offering suggestions and criticisms which have added materially to its interest.

The writer would also like to voice his appreciation to Ed Bierly whose magnificent illustrations adorn these pages. His is a talent with which it is a privilege to be associated.

Finally our thanks to the editor and his staff. It is not an easy task to combine text with illustrations, nor to match space with type, yet it has been done with feeling and precision.

Together, we hope that you will approve of our efforts. If through this booklet you gain a better understanding of the mammals that share the great outdoors with us, or if through it you should become aware of the urgent necessity of preserving some of our wild creatures, (and wild places), now before it is too late; we shall indeed be well repaid.

INTRODUCTION

Geographic Limitations

The only point in the United States at which four states adjoin is where Utah, Colorado, Arizona, and New Mexico come together. With adjacent portions of California, Nevada, and Texas, they contain all of our Southwestern Desert. Arizona and New Mexico especially, are known as desert States and for the most part deserve that appellation. Scattered over this desert country as though carelessly strewn by some giant hand are some of the highest and most beautiful mountains in our Nation. They may occur as isolated peaks magnificent in their loneliness, or as short ranges that continue but a little way before sinking to the level of the desert. On the other hand it is in Colorado that the Rocky Mountains reach their greatest height before merging with the high country in New Mexico, and all of the States mentioned have at least one range of major size.

Two great highways cross this area from East to West. U.S. 66, "Mainstreet of America," goes by way of Albuquerque and Flagstaff to Los Angeles; farther north U.S. 50 winds through the mountains from Pueblo to Salt Lake City and terminates at San Francisco. Significantly, they meet at St. Louis on their eastward course, and here for the moment we digress from geography to history.

Westward Ho

St. Louis in 1800 was a brawling frontier town. Strategically located at the point where the Missouri River meets the Mississippi, it was the jumping off place for those hardy souls adventurous enough to forsake the comforts of civilization for the unknown perils of the West. Already St. Louis was one of the fur centers of the world. Fashions of the day decreed that top hats be worn by men. The finest hats were made of beaver fur and no self-respecting dandy could be content with less. Trapping parties ascended the Missouri River as far as the mountains of Montana in search of pelts with which to supply the demand. When the animals became scarce in more accessible areas, trappers turned their attention to the mountains of New Mexico and Colorado. Hardships of the overland route, coupled with danger of attack by hostile Indians, discouraged all but the most hardy of a rugged breed. These "Mountain Men" as they became known, traveled in small parties with all the stealth and cunning of the Indians themselves. Gaunt from weeks of travel across the plains, they could rest in the Spanish settlement of Santa Fe for a few days before vanishing into the mountains. On the return trip they might again visit the Spanish pueblo or, eager for the night life of St. Louis, strike directly eastward across the prairies. Today's highways, while not following their trails directly, certainly parallel them to a great degree.

Little is known today of these early adventurers. A few written accounts have been printed, meager records of their catches have been noted, and here and there crude initials and dates carved on isolated canyon walls attest their passing. Their conquest of the West has faded into oblivion but it must be regarded as the opening wedge of American progress into the Southwest.

Mountains as Wildlife Reservoirs

Today's traveler spans in hours distances across these same routes that took weary weeks of heartbreaking toil a century ago. As he rides in cushioned ease he seldom pauses to reflect on the changes that have taken place since those early days. The great herds of bison with their attendant packs of wolves have vanished and in their place white-faced cattle graze on the level prairies. In the foothills the pronghorns have taken their last stand. Cities have sprung up on the camping sites of nomadic tribes that roamed the whole area between the Mississippi River and the Rocky Mountains. Only the mountains seem the same.

In winter these massive ranges form a barrier against the storms that sweep in from the northwest. More important—these great storehouses of our natural resources that in early days meant only gold and furs, and perhaps sudden death to the pioneers, have now been unlocked by their descendants. The glitter of gold and the glamour of furs pales when contrasted with the untold values that have since been taken out in the baser metals and lumber. This phase too is now coming to an end. It is becoming evident that in the face of our ever increasing population these natural playgrounds are destined to become a buffer against the tensions that we, as one of the most highly civilized peoples of the world, undergo in our daily life. Within another century they will represent one of the few remaining opportunities for many millions of Americans to get close to nature. As such the proper development and preservation of mountainous areas and their values is of vital importance to our Nation.

Mountains of the Southwestern States have been formed by three major agencies. These are, in order of importance, shrinkage of the earth's interior to form wrinkles on the surface; faulting, with subsequent erosion of exposed surfaces; and volcanic action. The first method is responsible for most of the large ranges, such as the coastal mountains of California and the Rocky Mountains. Faulting is responsible for many of the high plateau areas where one side may be a high rim or cliff and the other a gently sloping incline. The Mogollon Rim, extending across a part of Arizona and into New Mexico, is a classic example in this category. Volcanic action may result in great masses of igneous rock being extruded through cracks in the earth's surface or it may take the form of violent outbursts in one comparatively small area. Several mountain regions in Arizona and New Mexico are covered with huge fields of extruded lava. Capulin Mountain in New Mexico is an example of a recent volcano which built up an almost perfect cone of cinders and lava. Less noticeable than the mountains, but important nevertheless, are the tablelands of the Southwest. These mesas, too high to be typical of the desert, and in most cases too low to be considered as mountains, partake of the characteristics of both.

Desert "Islands"

The mountains of the Southwest have been compared to islands rising above the surface of a sea of desert. This is an apt comparison for not only do they differ materially from the hot, low desert in climate, but also in flora and fauna. Few species of either plants or animals living at these higher altitudes could survive conditions on the desert floor with any more success than land animals could take to the open sea. Their death from heat and aridity would only be more prolonged than that by drowning. Thus certain species isolated on mountain peaks are often as restricted in range as though they actually were surrounded by water. At times this results in such striking adaptation to local conditions that some common species become hardly recognizable. This is the exception to the rule however; most of the animals in this book are either of the same species as those in the Northern States or so closely allied that to the casual observer they will seem the same. Conditions that enable these species ordinarily associated with the snowy plains of the Midwest and the conifer forests of the North to live in the hot Southwest are brought about either directly or indirectly by altitude.

Life Zones

There are in this nucleus of four States a total of six life zones, (See map on page x.) The two lowest, the Lower and Upper Sonoran Life Zones, range from sea level to a maximum elevation of about 7000 feet. These two have been covered in the book "Mammals of the Southwest Deserts." The remaining four—Transition, Canadian, Hudsonian, and Alpine Life Zones—will furnish the material for this book. The names of these zones are self explanatory, because they are descriptive of those regions whose climates they approximate. Unlike the two life zones of the desert, which merge almost imperceptibly together, these upper zones are more sharply defined. They may often be identified at a great distance by their distinctive plant growth. It should be noted that plant species are even more susceptible to environmental factors than animals and are restricted to well defined areas within the extremes of temperature and moisture best suited to their individual needs. Thus each life zone has its typical plant species, and since animals in turn are dependent on certain plants for food or cover, one can often predict many of the species to be found in an individual area.

The Transition Life Zone in the Southwest usually lies at an altitude of between 7000 and 8000 feet. It encompasses the change from low trees and shrubs of the open desert to dense forest of the higher elevations. It is characterized by open forests of ponderosa pine usually intermingled with scattered thickets of Gambel oak. These trees are of a brighter green than the desert growth but do not compare with the deeper color of the firs that grow at a higher elevation.

The Canadian Life Zone begins at an altitude of about 8000 feet and extends to approximately 9500 feet. The Douglas-fir must be considered the outstanding species in this zone although the brilliant xiv autumn color of quaking aspens provides more spectacular identification of this area during the fall. Through the winter months when this tree has shed its leaves, the groves show up as gray patches among the dark green firs. At this elevation there is considerable snow during winter and correspondingly heavy rainfall in summer months. Under these favorable conditions there is usually a colorful display of wildflowers late in the spring.

The Hudsonian Life Zone is marked by a noticeable decrease in numbers of plant species. At this altitude, (9500 to 11,500 feet), the winters are severe and summers of short duration. This is the zone of white fir which grows tall and slim so to better shed its seasonal burden of snow and sleet. In the more sheltered places spruce finds a habitat suited to its needs. Near the upper edge of the Hudsonian Life Zone the trees become stunted and misshapen and finally disappear entirely. This is timberline; the beginning of the Alpine Life Zone, or as it often called, the Arctic-Alpine Life Zone.

Here is a world of barren rock and biting cold. At 12,000 feet and above the eternal snows lie deep on the peaks. Yet, even though at first glance there seems to be little evidence of life of any kind, a close scrutiny will reveal low mat-like plants growing among the exposed rocks and tiny paths leading to burrows in the rock slides. Among the larger mammals there are few other than the mountain sheep that can endure the rigors of this inhospitable region.

These are the zones of the Southwest uplands. Altitudes given are approximate and apply to such mountain ranges as the San Francisco Peaks of Arizona and the Sangre de Cristo Mountains in New Mexico. As one travels farther north the zones descend ever lower until in the Far North the Arctic-Alpine Life Zone is found at sea level. Since climate more than any other factor, determines the types of plants and animals that can live in a given area it is only natural that on these mountain islands many species entirely foreign to the surrounding deserts are found at home. Though it would seem that because of the relative abundance of water at higher elevations the upland species would have the better environment, this is not entirely true. Balanced against this advantage are the severe winters which, in addition to freezing temperatures, usher in a period of deep snows and famine. Even though many species show a high degree of adaptation to these conditions, an especially long or cold winter season will result in the death of weaker individuals.

Man and Wilderness

The effects of man's presence on the upland species is perhaps not as serious as on those of the desert. Though he has been instrumental in upsetting the balance of nature everywhere, it has been chiefly through agriculture and grazing. Because of the rough broken character of much high country in the Southwest the first is impossible in many cases and the second only partially successful. There are other factors however which menace the future of the upland species. Among these are: hunting pressures, predator control, and lumbering. Even fire control, admirable as it may be for human purposes, disrupts the long cycles which are a normal part of plant and animal succession in forested areas. These are only a few of the means by which man deliberately or unconsciously decimates the animal population. They are set down as a reminder that unless conservation and science cooperate in management problems, it is conceivable that many of our common species could well become extinct within the next 100 years. Our natural resources are our heritage; let us not waste the substance of our trust.

As our wilderness areas shrink it seems that our people are gradually becoming more interested, not only in the welfare of our native species but in their ways as well. This type of curiosity augurs well for the future of our remaining wild creatures. In times past an interest in mammals was limited mainly to sportsmen who often knew a great deal about where to find game animals and how to pursue them. Their interest usually ended with the shot that brought the quarry down. Today many people have discovered that a study of the habits of any animal in its native habitat is a fascinating out-of-doors hobby in a virtually untouched field. With patience and attention to details the layman will occasionally discover facts about the daily life of some common species that have escaped the attention of our foremost naturalists. This is no criticism of the scientific approach. It is recommended that for his own benefit the nature enthusiast learn a few of the fundamentals of zoology, particularly of classification and taxonomy, which mean the grouping and naming of species.

Classification of Animals

Classification of animals is easy to understand. Briefly, they are divided into large groups called *orders*. These are further divided into *genera*, and the genera in turn contain one or more *species*.

Scientific names of animals are always given in Latin. Written in this universal language they are intelligible to all scientists, regardless of nationality. It is a mistake to shy away from them because they are cumbersome and unfamiliar to the eye. They usually reveal some important characteristic of the animal they stand for. This is their true function; it seems to this writer that it is a mistake to name an animal after a geographical location or a person, although it is frequently done. The literal translations of specific names in this book will illustrate this point. See how much more interesting and how much more easily remembered those names are which describe habits or physical attributes of the creature.

Described herein are but a part of the species native to the Southwestern uplands. Those chosen were selected because they are either common, rare, or unusually interesting. Collectively they make up a representative cross section of the mammals that live above the deserts of the Southwest.

For further information on these and other mammals of the region see the list of references on page 123.

HOOFED ANIMALS
Artiodactyla
(even-toed hoofed animals)

This order includes all of the hoofed animals native to the United States. These are the mammals which are ordinarily spoken of as the "cloven-hoofed animals." An odd-toed group (*Perissodactyla*), which includes the so-called wild horses and burros, cannot properly be included as natives since these animals date back only to the time of the Spanish conquest of our Southwest. In earlier geologic ages horses ranged this continent, but in more primitive forms than those now found in other parts of the world.

Through a study of fossil forms it has been determined that our present hoofed animals evolved from creatures which lived on the edges of the great tropical swamps that once covered large areas of our present land masses. They were long-legged and splay-footed, well adapted to an environment of deep mud and lush vegetation. As the waters gradually disappeared and the animals were forced to take to dry land, their strange feet underwent a slow transformation. Because they had become accustomed to walking on the tips of their toes to stay up out of the mud, the first toe did not touch solid ground at all in this new environment. Since it was of no use it soon vanished entirely or became vestigial. Some species developed a divided foot in which the second and third toes and the fourth and fifth toes combined respectively to bear the animal's weight. Eventually the third and fourth toes assumed this responsibility alone, and the second and fifth toes became dew claws. These are the cloven-hoofed animals of today. In other species the third toe was developed to bear the weight, and this resulted in a single-toed group of which the horse is an example. In all cases an enormous modification of the nails or claws with which most animals are equipped has resulted in that protective covering called the hoof. The under surface of the foot is somewhat softer and corresponds to the heavy pad that protects the bottom of a dog's toe. This brief explanation refers only in the broadest sense to the order as represented in the United States. The feet of the various species have become so specialized to their separate ways of life that an individual can usually be easily identified by its tracks alone. It is quite possible that many species are still undergoing subtle changes in this respect.

With but one exception the cloven-hoofed animals of our southwestern mountains bear either horns or antlers. The exception is the collared peccary, "javelina," (*pecari tajacu*) which, during the heat of the summer, sometimes ascends to the Transition Life Zone in southern Arizona and southwestern New Mexico. Essentially an animal of the low desert, it will not be included in this book. The species which have hollow, permanent horns are the bighorn and pronghorn. The pronghorn is distinctive in shedding the sheaths of its horns each year, but the hollow, bony core remains intact. In this group both sexes bear horns. Animals bearing antlers are the elk and the deer. The antlers are deciduous, being shed each year at about the same time as the winter coat. Only the males of these species have antlers, any female with antlers can be considered abnormal.

The Southwest is fortunate in still having a number of the species of this order native to the United States. The bison can hardly be considered a wild species, since it exists now only through the efforts of a few conservationists who brought it back from virtual extinction. Mountain goats, caribou, and moose are the only other species not known to inhabit the Southwest.

In Nature's balance the order *Artiodactyla* seems to have been meant as food for the large predators. Their protection against the flesh eaters consists mainly in fleetness of foot, keen hearing, and a wide range of vision, as evidenced by the large eyes set in the sides of the head. They are but poorly equipped to actively resist attack by the larger carnivores. Their best defense is flight.

Bighorn (mountain sheep)
Ovis canadensis (Latin: a sheep from Canada)

RANGE: This species, with its several varieties, inhabits most of the mountainous region of the western United States. In Mexico it occurs in the northern Sierra Madres and over almost the whole length of Baja California.

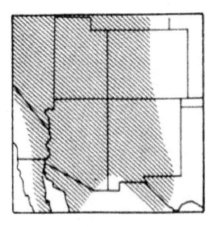

HABITAT: Among or in the vicinity of more precipitous places in the mountains.

DESCRIPTION: A blocky animal, rather large, with heavy, curving horns. Total length of adult male 5 feet. Tail about 5 inches. Weight up to 275 pounds. General color a dark gray to brown with lighter areas underneath belly and inside of legs. The rump patch is much lighter than any other part of the body; in most cases it can be described as white. Females are similar in appearance to the males except that they are smaller and the horns are much shorter and slimmer. Young, one or two, twins being common.

Interesting as the desert varieties of this species may be in their adaptation to an environment that seems foreign to their nature, they cannot compare with the high mountain animal. Seen against the backdrop of a great gray cliff or silhouetted against the skyline of a snowy crest the bighorn has a magnificence that is thrilling. In flight it is even more spectacular as it bounds from one narrow shelf to another in an incredible show of surefootedness. Yet this airy grace is exhibited by a chunky animal that often weighs well over 200 pounds. The secret lies in the specially adapted hooves with bottoms that cling to smooth surfaces like crepe rubber and edges that cut into snow and ice or gain a purchase on the smallest projections of the rocks. The legs and body, though heavy, are well proportioned and so extremely well muscled that no matter what demands are placed on them this sheep seems to have a comfortable reserve of power. No doubt the display of complete coordination adds to the illusion of ease with which it ascends to the most inaccessible places. Descents often are even more spectacular, the animal seldom hesitating at vertical leaps of 15 feet or more down from one narrow ledge to another.

bighorn

In the high mountains where this sheep prefers to make its home it usually ranges near or above timberline. During winter storms it may sometimes be forced down into the shelter of the forests, but as soon as conditions warrant it will go back to its world of barren rocks and snow. Here, with an unobstructed view, its keen eyes can pick out the stealthy movements of the mountain lion, the only mammal predator capable of making any serious inroads on its numbers. It has few other natural enemies. A golden eagle occasionally may strike a lamb and knock it from a ledge, or a high ranging bobcat or lynx may be lucky enough to snatch a very young one away from its mother, but these are rare occurrences.

Bighorns depend mainly on browse for food. This is only natural since in the high altitudes they frequent little grass can be found. Usually there is some abundance of low shrubs growing in crevices on the rocks, however, and many of the tiny annuals are also searched out during the short summer season. At times a sheltered cove on the south exposure of a mountain will become filled with such shrubs as the snowberry, and the sheep take full advantage of such situations. As a rule they keep well fed for, scanty though it seems, there are few competitors for the food supply above timberline.

I have observed these sheep many times in the Rockies. Perhaps my most memorable experience with this species was on Mount Cochran in southern Montana. It was a gray, blustery day in September with occasional snow flurries sweeping about the summits. On the eastern exposure of the mountain a steep slope of slide rock extended for perhaps 1,000 feet from one of the upper ridges to timberline. Not expecting to see any game at that elevation, I made my way up this slope with no effort to keep quiet. In my progress several rocks were dislodged and went rattling down across the mass of talus. At the summit of the ridge a low escarpment made a convenient windbreak against the gale that was tearing the clouds to shreds as they came drifting up the opposite slope, and I sat down to catch my breath before entering its full force. As I sat there surveying the scene spread out below, my attention was attracted by a low cough close by. Looking to the left about 40 feet away and 15 feet above me, I saw two magnificent rams standing on a projecting point looking down at me. They seemed to have no fear; rather they evinced a deep curiosity as to what strange animal this was that had wandered into their domain. For the better part of a half hour I hardly dared breathe for fear of frightening them. At first they gazed at me fixedly, occasionally giving a low snorting cough and stamping their feet. Then as I did not move, they would wheel about and change positions, sometimes taking a long look over the mountains before bringing their attention back. Finally when the cold had penetrated to my very bones, I stood up. They were away in a flash, reappearing from behind their vantage point with two ewes and an almost full-grown lamb following them. While I watched they dashed at a sheer cliff that reared up to the summit, and with hardly slackening speed bounded up its face until they were lost in the clouds.

Although this happened in 1928 the experience is as vivid in my mind as though it happened yesterday. Perhaps the most striking feature of bighorns seen at this close range is the eyes. They might be described as a clear, golden amber with a long oval, velvety black pupil. Credited with telescopic vision, they must be some of the most useful as well as beautiful eyes to be found in the animal kingdom.

Pronghorn (antelope)
Antilocapra americana (Latin: antelope and goat, American)

RANGE: West Texas, eastern Colorado and central Wyoming to southern California and western Nevada, and from southern Saskatchewan south into northern Mexico.

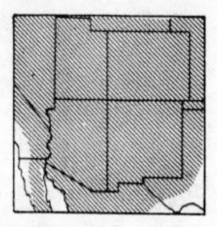

HABITAT: Grasslands of mesas and prairies, mostly in the Upper Sonoran Zone.

DESCRIPTION: A white and tan colored animal, considerably smaller than a deer; horns with a single flat prong curving forward. Total length about 4 feet. Tail about 6 inches. Average weight 100 to 125 pounds. Color, tan or black shading to white under belly and insides of legs. Two conspicuous white bands under the neck, and the large white rump patch of erectile hairs are unlike the markings of any other native animal. A short, stiff mane of dark hairs follows the back of the neck from ears to shoulders. Hooves black, horns also black except for the light tips on those of older males. Both sexes horned. Young, usually two, born in May.

pronghorn

Pronghorns are unique among cloven-hoofed animals of the Southwest. There is only one species, with several subspecies; a variety *mexicana*, once common along the Mexican border, is considered extinct in this country. The pronghorn has no "dew claws" like most other animals with divided hooves. It has permanent bony cores in its horns but sheds the outer sheaths each year. When these drop off the succeeding sheaths are already well developed. Although at first these new sheaths are soft and covered with a scanty growth of short stiff hairs, corresponding to the velvet in antlered animals, it does not take long for them to harden and become dangerous weapons. They reach full development at about the time of the rut; bucks have been known to fight to the death in the savage encounters that break out at this time.

Were it not for its unusual horns the pronghorn probably would be known by a common name such as the white-tailed antelope, for the beautiful white rump patch is undoubtedly its next most conspicuous feature. However, at least two other animals have been named "antelope" because their posteriors have some similarity. They are the white-tailed ground squirrel and the antelope jackrabbit of the Sonoran Life Zone. The ground squirrel (*Citellus leucurus*) has merely a white ventral surface on its tail which may or may not act as a flashing signal when flipped about, but the antelope jackrabbit (*Lepus alleni*) has a rump patch that bears a striking likeness to the pronghorn's both in appearance and manner of use. In both cases the rump patches are composed of long, erectile white hairs which are raised when the animal is alarmed. In flight they are thought to act as warning signals; at any rate they are very effective in catching the eye, and on the open plains the pronghorn's can be seen at a distance where the rest of the animal is indistinguishable. It may well be, on the other hand, that this flashy ornament is meant to attract the attention of an enemy and lead it in pursuit of an adult individual rather than allowing it to discover the helpless young. Neither animal can be matched in speed on level ground by any native four-footed predator.

In times past the pronghorn usually lived in the valley and prairie country. In the Southwest it roamed over much of both the Upper and Lower Sonoran Life Zones, wherever suitable grass and herbage could be found. On the prairies of the Midwest bands of pronghorns grazed in close proximity to herds of buffalo. During the middle of the last century it was the only animal whose numbers even approached those of the latter. More adaptable than the buffalo, it has retreated before the advance of civilization and taken up new ranges in rough and broken country which is unsuited to agriculture. As a rule this is much higher than its former range. Pronghorns are no longer found in the Lower Sonoran Life Zone, except as small bands that have been introduced from farther north. The greatest population now ranges in the upper portions of the Upper Sonoran and along the lower fringe of the Transition Life Zone. The animal has always been considered migratory to some extent because it moved from mesa summer ranges to the protection of warmer valleys during winter months. This habit is even more pronounced in later years at the higher levels it now inhabits. These slim, long-legged creatures are virtually helpless in deep snow and avoid it whenever possible. They have even been known to mingle with cattle and join with them at the feed racks during severe winters, an indication of the extreme need to which shy pronghorns are sometimes reduced.

They are essentially grazing animals. In the past they ate prairie grasses during the summer; in winter these same grasses made excellent hay that lost little in nourishment from having dried on the roots. In addition, they ate low herbage and nibbled leaves, buds, and fruits from shrubs that grew along the watercourses. Their food today is much the same except that in the many areas where they receive competition from range cattle they undoubtedly resort to more browse than formerly.

Natural enemies of the pronghorn are legion, their success indifferent. Every large carnivore will snap at the chance to take one, and even the golden eagle has been known to kill them. The most serious depredations are carried out on those young too small to follow the mother. However, these attacks are fraught with danger, for the females are very courageous in the defense of their young and at times several will join in routing an enemy. In addition to this protection accorded them by adult members of the band, the young have an almost perfect camouflage in their plain coats that blend so closely with the color of the grass in which they usually lie concealed. Because of their fleetness, few adults fall prey to predators. Many attempts have been made to clock the speed of the pronghorn in full flight but the estimates vary greatly. Although a fast horse can keep up with one on smooth, level ground, it is soon outdistanced on stony soil or in rough country.

baby pronghorn

Bison (buffalo)
***Bison bison* (Teutonic name given to this animal)**

RANGE: At present bison exist only in widely scattered sanctuaries. In Colonial times they ranged from southern Alaska to the Texas plains, from the Rocky Mountains to the Atlantic, and as far south as Georgia. They are known in historic times in Utah, Colorado, and New Mexico.

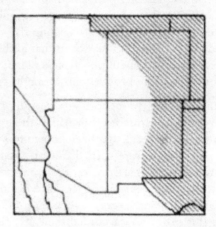

HABITAT: Mainly grasslands; a comparatively small number known locally as "wood" bison lived in the fringes of the forests.

DESCRIPTION: Although bison are familiar to almost everyone, some figures on weights and dimensions may be surprising. Bulls weigh up to 1800 lbs., reach 6 feet height at the shoulders and up to 11 feet in length, of which about 2 feet is the short tail. Cows average much smaller from 800 to 1000 pounds, and rarely over 7 feet long. Both sexes have heavy, black, sharply curving horns, tapering quickly to a point, and a heavy growth of woolly hair covering most of the head and forequarters. A large hump lies over the latter and descends sharply to the neck. The head is massive, horns widely spaced, and small eyes set far apart. A heavy "goatee" swings from the lower jaw. All these features combine to give the animal a most ponderous appearance. Nevertheless, bison are surprisingly agile and are not creatures with which to trifle, especially in the breeding season, when bulls will charge with little provocation. Like most wild cattle, bison normally bear but one calf per year. Twins are uncommon.

The history of the bison is unique in the annals of American mammalogy. It hinges on simple economics, reflecting transfer of the western prairies from Indian control to white. It is a pitifully short history in its final stages, requiring only 50 years to drive a massive species, numbering in the millions, from a well balanced existence to near extinction. Yet considering the nature of civilization and progress there could have been no other end, so perhaps it is well that it was quickly over.

For countless centuries the bison had roamed the prairies, their seasonal migrations making eddies in the great herds that darkened the plains. They were host to the Indian, and to the gray wolf, yet so well were they adapted to their life that these depredations were merely normal inroads on their numbers. They drifted with the seasons and the weather cycles, grazing on the nutritious grasses of the prairie. Weather and food supply; these were the main factors which controlled the "buffalo" population until the coming of the white man.

The first white man to see an American bison is thought to have been Cortez, who in 1521 wrote of such an animal in Montezuma's collection of animals. This menagerie was kept in the Aztec capitol on the site of what is now Mexico City. There the bison was an exotic, hundreds of miles south of its range. In 1540 Coronado found the Zuni Indians in northern New Mexico using bison hides, and a short distance northeast of that point encountered the species on the great plains. The eastern edge of the Rio Grande Valley in New Mexico seems to have been the western limit of bison in the Southwest. Unfavorable climate, plus the comparatively heavy Indian population of the valley probably combined to halt farther penetration in that direction.

From 1540 until 1840 the white man limited his activities on the western plains to exploring. American colonization had reached the Mississippi River, but remained there while gathering its forces for the expansion which later settled the West. Under Mexican rule, the Southwest progressed very slowly. Then in the span of 50 years a chain of events occurred which determined the destiny of the West and sealed the fate of the bison herd.

bison

Outstanding among these events were: the War with Mexico, 1844; the 1849 Gold Rush to California; the Gadsden Purchase in 1854; and completion of the transcontinental railroad in 1868. The first three added new and important territory to the United States. This made construction of transportation and communication facilities a vital necessity, hence the railroad. Completion of the Union Pacific Railroad in 1868 divided the bison population into southern and northern herds and made market hunting profitable. Three factors contributed to extermination: profit in the traffic of hides, meat, and bones; control of troublesome Indian tribes through elimination of one of their major sources of food; and finally, removal of any competition on the grassy plains of Texas and Kansas against the great herds of Longhorn cattle which were beginning to make Western range history. In 1874, only 6 years after completion of the Union Pacific, the slaughter of the southern herd was complete. It is of interest to note that not one piece of legislation was passed to protect the southern herd.

The northern herd did slightly better. Closed seasons were established in Idaho in 1864, in Wyoming in 1871, Montana in 1872, Nebraska in 1875, Colorado in 1877, New Mexico in 1880, North and South Dakota in 1883. Nevertheless, the herd dwindled, and by 1890 was nearly extinct. Since that time, through careful management, a few small herds have been established in Parks and refuges, but today the bison must be considered more a domesticated animal than a wild one.

Although the animal was not as important economically to the southwestern as to the plains Indians, it was a religious symbol of some value. Archeological finds far west of the historic range, and dances still used in ceremonies, reveal that several southwestern tribes sent hunting parties eastward into bison country. This must have been very dangerous, for plains Indians would have considered them invaders. Bison were food, shelter, and clothing to them. Imagine their consternation when white men began to slaughter the source of their living.

There are today but few reminders of the great herds of the west. Perhaps one well versed in the ways of these wild cattle could still find traces of the deeply cut trails which led to the watering places, or shallow depressions where the clumsy beasts once wallowed in the mud. Many of the Indian dances recall the importance of this animal to primitive man. Perhaps our most constant reminder is the coin which commemorates this symbol of the wild west, showing the Indian on one side and the bison on the other.

Mule deer
Odocoileus hemionus (Greek: odous, tooth and koilus, hollow. Greek: hemionus, mule)

RANGE: Western half of North America from Central Canada to central Mexico.

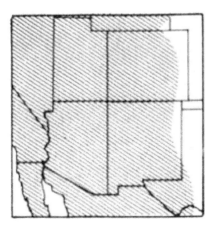

HABITAT: Forests and brushy areas from near sea level to lower edge of the Alpine Life Zone.

DESCRIPTION: A large-eared deer with a tail that is either all black above or black tipped. Total length of an average adult about 6 feet. Tail about 8 inches. The coat is reddish in summer and blue-gray in winter. Under parts and insides of legs lighter in color. Some forms of this species have a white rump patch, others none. The tail may be black-tipped, or black over the whole dorsal surface, but is more sparsely haired than that of other native deer and is naked over at least part of the under surface. Only the bucks have antlers. These are typical in forking equally from the main beam. They are shed every year.

mule deer

The mule deer is typical of the western mountains. Even in early days it was never found east of the Mississippi and now is seldom seen east of the Rockies. Only one species is recognized in the United States, although over its vast range are many subspecific forms. All are notable for the size of their ears, from which derives the common name "mule." The black-tailed deer of the Pacific Coast, long considered a distinct species, is now rated a subspecies of the mule deer.

In a general way the deer of the United States may be divided into two groups, these separated geographically by the Continental Divide. East of this line is the territory occupied by the white-tailed group; westward of it live the mule deer. Inasmuch as species seldom stop abruptly at geographical lines, we find in this instance that a whitetail subspecies, locally known as the Sonora fantail, is found along the Mexican border as far west as the Colorado River, territory also occupied by desert-dwelling mule deer. In like fashion the mule deer of the Rocky Mountains can even now be seen in the Badlands of North Dakota, several hundred miles east of the Continental Divide and well within the western range of the plains white-tailed.

Though the two species mingle in places, they are easily distinguished from each other, even by the novice. Because in many cases the animal is seen only in flight, the manner of running is perhaps the most prominent field difference. The mule deer, adapted to living in rough country, bounds away in stiff-legged jumps that look rather awkward on the level but can carry it up a steep incline with surprising speed. The white-tailed, on the other hand, stretches out and runs at a bobbing gallop. Deer seldom take leisurely flight from a human, usually straining every muscle to leave their enemy as soon as possible. In the rough, broken country frequented by mule deer this tactic often makes considerable commotion.

I am reminded of a herd of an estimated 70 deer that I jumped on a steep mountainside in southern Utah. The crashing of brush, crackling of hooves, and noise of rocks kicked loose in their flight created the impression of a landslide.

Another easily seen field difference between mule and white-tailed deer is the dark, short-haired tail of the former as compared with the great white fan of the latter. The tail of the mule deer seems in no way to be used as a signal. In flight it is not wagged from side to side as is that of the white-tailed.

Antlers of the mule deer are unlike those of any other large game species inhabiting their range. They are impressively large as a rule and, because of the high angle at which they rise from the head, often look larger. The spread is wide in proportion to the height; thus it is not unusual for a well-antlered buck to be mistaken for an elk, especially at a distance. The antlers are unique in having a beam that forks equally to form the points. Thus a typical head might have five points, these being: the basal snag, a small tine rising from the beam near the head; and four points, two from the forking of each division of the beam. The western manner of counting the points consists of numbering those of one antler only; the method often used in the East counts all of the points of both. The number of points does not necessarily denote the age of a deer. Under normal conditions the antlers will increase in size and points with every new pair until maturity is attained. They will then grow to approximately the same size for several years. In old age, the antler development will usually dwindle with each succeeding year until, in senility, they may be as small as those of a young deer. The condition of teeth and hooves is a more accurate indication of age even though this method lacks prestige of the time-honored system of points.

It would seem, from the ease with which this big deer can be established in varying types of habitat, that it is in little danger of extinction. It is probable that the various subspecies will disappear before long because their range is rapidly being taken up by agriculture or lumbering. Given some protection, the species will endure in the higher mountains for many years to come.

White-tailed deer
***Odocoileus virginianus* (Greek: odous, tooth and koilus, hollow. Latin: of Virginia)**

white-tailed deer

RANGE: Mostly east of the Continental Divide in the United States, north into southern Canada, and most of Mexico except Baja California.

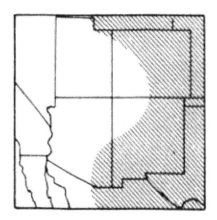

HABITAT: Brushy and wooded country.

DESCRIPTION: A deer with a large, white tail, held aloft and wagged from side to side as it runs away through the underbrush. In the Southwest, two geographic variants occur, the subspecies *virginianus* and the subspecies *couesi*; the latter known locally as Sonora fantail, and seen in the United States only in a limited range along the border. *Odocoileus virginianus* of the Southwest is a large deer. It usually weighs between 150 and 250 pounds, and sometimes up to 300. The average adult animal will measure around 6 feet in total length. Tail about 10 inches. Color is reddish in summer, changing to gray with the winter coat. Belly, insides of legs, and undersurface of tail are white. Ears are small. Antlers have upright tines from a single beam.

As the specific name indicates, this is the same deer that is found in the Eastern States. It is also known as the plains whitetail, because it was once common along brushy draws and river bottoms throughout the prairie regions. Preeminently an eastern animal, it occurs most abundantly in the Eastern States, dwindling in numbers westward to the Continental Divide. A few scattered groups are found in the Pacific Northwest, and the subspecies *couesi* extends westward along the Mexican border to the Colorado River.

The white-tailed deer may be distinguished from mule deer by any of three characteristics, all readily apparent in the field. These are: shape and construction of antlers, size and color of tail, and method of running. Antlers consist of two main beams which, after rising from the head, curve forward almost at right angles with a line drawn from forehead to nose. The tines rise from these main beams. In the mule deer the beams rise at a higher angle from the head and fork rather than remain single. The white-tailed tail is long and bushy, fully haired all around and pure white beneath. In flight it is erected and "wigwagged" from side to side. This, together with the white insides of the hams, presents a great show of white hair as the animal retreats. The mule deer has a thin, sparsely-haired tail that is bare underneath and does not wave from side to side in running. The "whitetail" runs at a brisk gallop with belly close to the ground; the mule deer bounds away with a series of ballet-like leaps.

This is the deer that contributed so much to the pioneers in their westward trek from the Atlantic States. It was important not only for its flesh but for its hide, which after tanning became the buckskin moccasins, breeches, and coats commonly worn by outdoorsmen in early days. Its distribution is now spotty compared with the former range, although there are today probably more white-tailed deer in the United States than in colonial times. This is mainly because in the thickly settled Eastern States predators have been reduced to a minimum and hunting seasons carefully regulated. It is too early yet to know if predator elimination will result in an inferior strain of deer, but the relative overpopulation in many localities has been indicated by lack of browse, disease, and excessive winter kill. The latter especially has been a problem in some of the Northern States. "Whitetails" are gregarious creatures, banding together in considerable numbers at times, especially during winter. A band of them in deep snow will stay together and their hooves will soon tramp down the snow over a small area. As succeeding snows fall, the drifts become deeper around the "deer yards" and eventually the occupants become as imprisoned by this white barrier as though they were fenced. If the number of animals in the yard is too great, available browse soon disappears and many will starve to death before warm weather returns. Over most of the mountainous area occupied by white-tailed deer in the Southwest snow is no problem. The herds merely move down to lower country when the snow gets too deep. This seasonal movement is so pronounced that this deer is classed as a migratory animal in some localities.

In line with this migratory tendency the "whitetail" follows a varied but well-marked routine in its life pattern. About the time of shedding the winter coat late in the spring, the bucks also cast their antlers. With the loss of these beautiful weapons their personalities suffer. They leave the group with which they have spent the winter and ascend to the higher mountains, there to consort with a few similarly afflicted individuals until a new growth of antlers restores their dignity. The does, left behind, have problems of their own. These include driving the yearling fawns away to fend for themselves in order that the does may give undivided attention to the tiny, spotted newcomers that arrive in midsummer. By this time the adults have put on the short, yellowish-red summer coat. The fawns are reddish too, but covered with pale spots, a combination that blends well with lights and shadows in the brushy places where the does choose to hide them. As soon as the fawns are large enough to follow their mothers the little family groups begin a gradual trek up the mountainside. There are several reasons for this exodus, chief of which are cooler temperatures, better browse, and fewer stinging insects.

While the does have been rearing their young, the bucks have been staging a slow comeback on the ridges above. By early fall their new antlers have hardened, been cleaned of velvet, and polished in the brushy thickets. With restored weapons they again seek company of the does. The season of the rut comes in a time when the bucks are at the peak of vigor and combativeness. Yearlings and weaker bucks are soon outclassed, leaving the most virile and aggressive males to become progenitors of the following year's fawns. The simplicity of this system is equalled only by its effectiveness. Natural selective breeding is one of the most important items in perpetuation of a species. A decline in numbers of the best breeding animals often results in an inferior strain. In conservation of deer herds it is well to remember that it is not always the *number* of animals that is the prime consideration. A smaller group of healthy, vigorous individuals is usually more to be desired than a larger population in average condition.

Although the species has vanished from many of its haunts in the Prairie States, it will not likely become extinct for a long time. Ranked by many authorities as our foremost game animal, it has been the "guinea pig" in many conservation experiments. Adaptable to almost any environment with suitable shelter and browse factors, it needs only a little protection to become well established. The "key" deer of the Florida Everglades, a tiny animal attaining a weight of only 50 pounds, is, however, on the verge of extinction. Another subspecies, the "Sonora fantail," native to Mexico and the southwestern United States, is greatly reduced in numbers and seems destined to vanish.

Elk
Cervus canadensis (Latin: stag or deer, from Canada)

RANGE: Along the Rocky Mountains of the United States and Canada. Also found in central Canada, western Oregon and Washington, central California, and various small areas in those Western States where it has been introduced.

HABITAT: Wooded places and high sheltered mountain valleys.

DESCRIPTION: A very large deer with enormous antlers, a thin neck, and a light rump patch. Total length 80 to 100 inches. Tail 4 to 5 inches. Shoulder height 49 to 59 inches. Average weight 600 to 700 pounds, with a maximum of about 1100. Coat dark in summer, lighter in winter. Longer hair on neck and throat of the bull forms a mane that is distinguishable at some distance. Antlers extremely large, usually six points on adult males. Females do not normally bear antlers. Hooves are black. Young usually one, although twins not rare.

The elk is the largest member of the deer family native to the southwestern United States. It was once widely distributed, known not only throughout the Middle West but also in most of the Eastern States. In fact, one of its common names, "wapiti," is of eastern American Indian origin; it was so called by the Algonquins. The Crees of Canada and the Northern States call it "wapitiu" (pale white) to distinguish it from the darker colored moose with which it was associated in that region. It is now confined to the Rockies and westward in the United States, and to the Rockies and central portion of Canada. Many herds now found in Western States have been introduced to take the place of those thoughtlessly exterminated in the early days. This has been the case in Arizona and New Mexico, where Merriam's elk disappeared before 1900. This elk, now known only from scanty records and a few mounted heads and skulls, was a giant of its kind. Not only was it larger than the Wyoming elk which now takes its place, but it had correspondingly massive antlers. Its passing is a grim warning of what could happen easily to the tule elk, a pygmy elk of central California which has been reduced to a dangerously small herd. The elk now present in the Southwest, chiefly if not all, are descendants of individuals brought down from the large herds of the Yellowstone Park area. In their new homeland they maintain the same habits that characterize the species in Wyoming.

Next to buffalo, elk are the most gregarious large mammals in the United States. The degree to which they band together varies with the seasons and can be attributed to their migratory instincts. During summer months the bands are small and widely scattered high in the Transition Life Zone and even higher at times. With the advent of cold weather they work their way down to lower country, and winter finds them gathered in sheltered grassy valleys. This exodus to winter quarters can be one of the most thrilling sights in Nature. In the north it is not uncommon for herds of a thousand or more of these stately animals to move into one of the more favored valleys. They have the instinct so highly developed among most animals of knowing when a storm is imminent, and the migration may be completed within a period of 48 hours, or even less if foul weather is brewing.

The concentration of hundreds of these hungry animals into one small area creates numerous problems, the most serious being that of feed. Before the white man came, the elk population was more scattered, and many winter feeding grounds were available. In those ungrazed areas they were able to paw down through the snow to the nourishing dry grass beneath. Large herds must now be fed on hay to avoid winter losses that would otherwise result. In the Southwest, with its comparatively mild winters and small population, the animals experience little difficulty in weathering the storms without human aid. The present herds appear well established, and with proper conservation measures should be a valuable part of our wildlife for many years to come.

elk

Migratory though they are, elk still must weather a great seasonal range of temperature. In adapting to these changes they have developed two definite coats, one for summer and one for winter. The winter garb is put on early in the fall; it consists of a heavy coat of brown woolly underfur with guard hairs that vary from gray on the sides to almost black on neck and legs. Old bulls tend to be more black and white than the cows and younger animals. This heavy pelage, often called the "gray" coat, effectively wards off cold winds that sweep through the mountains and insulates the wearer against snow that is driven into the outer surface. In the spring this coat is shed to make way for a light summer coat. The matted hair falls away in great bunches, and the animals are unkempt in appearance for 2 months or more. The summer coat is made up of short, stiff hairs with little underfur. The pelage is glossy when compared with the harsh guard hairs of the winter coat. In color it is tawny, appearing reddish at a distance. The rump patch is a light tawny color in both coats.

With the coming of spring the bulls lose the great antlers which they have carried through the winter. This takes place through a general deterioration of the antler base accompanied by some reabsorption of tissues at that point. The antlers may simply drop off or, in their weakened condition, be snapped off on contact with low hanging branches. They are usually shed in March, and by May a new pair begins to grow. As with the rest of the deer family, a thick growth of velvet covers the new growth. The first stages look rather ludicrous as the antlers develop points by successive stages, each tine coming to maturity before the next begins to grow. Eventually the height of the antlers "catches up," so to speak, with the overprominent base. At full maturity, attained by August, there are few sights so impressive as a bull elk in the velvet. When this stage is reached the antlers, until now extremely tender, begin to harden and lose their sense of feeling. The bulls strip off the velvet by rubbing against branches and brush. Gradually the hard core emerges, stained a rich brown, except for the tips of the tines which are a gleaming ivory white. The antlers are so beautifully symmetrical that they seem graceful despite their size. One of the largest pairs on record has a length of beam of 64¾ inches and a spread of 74 inches.

A mature bull usually has six tines on each antler. These have definite names. The first tine extends forward from the head and is known as the "brow" tine; the next to it as the "bez" tine. Collectively they are called the "lifters," formerly known as "war tines." The next point inclines toward the vertical; this is the "trez" tine. The fourth is the "royal" or "dagger" point, and the terminal fork of the antler forms the final two points which are called "surroyals."

Unwieldy as this tremendous rack of antlers appears, the animals handle them with comparative ease. In the normal walk or trot the body is carried along smoothly with the nose held up and forward. In this posture the antlers are well balanced and are carried without undue strain. In running through brush the nose is lifted still higher; this throws the antlers farther back along the shoulders, and as the nose parts the branches they slide along the curving beams without catching on the tines. Despite these cumbersome impediments, the elk creates less disturbance than most large forest animals when in flight. Antlers as weapons of offense are far overrated, for they seldom serve this function. Males have been severely injured and even killed in fights among themselves, but these are exceptions, and most fighting is done by striking with the front feet. If antlers are used it is usually with a chopping, downward motion that rakes, rather than puncturing the hide of the opponent.

Despite the fine appearance he presents, the bull elk is not content merely to be seen, but insists on being heard as well. His vocal effort is a high, clear, mellow tone commonly known as bugling, although it seems to have more the quality of a whistle than the sound of a horn. The call begins on a low note that is sustained for perhaps two seconds and then rises swiftly for a full octave to a sweet mellow crescendo, drops by swift degrees to the first note, and dies away. This is followed by several coughing grunts that can be heard only at close range. Bugling can be heard for a great distance, and on a clear quiet evening one of the greatest charms of wilderness camping is to hear this clear challenge flung out from some nearby ridge. The response is quickly returned from other hillsides, some so far away as to be mere whispers in the distance.

Bugling is indulged mainly during the rutting season and lasts from August to November. During this time it undoubtedly is intended as a challenge to other bulls and perhaps also to impress the cows with their lords' great importance. At other seasons it is heard but infrequently, and then probably is simply an expression of abundant animal spirits. Cows have been known to bugle, but this is a rare occurrence.

The single calf is born between mid-May and mid-June. Twins are not uncommon. At birth the calf will weigh 30 to 40 pounds, and is an awkward animal. It has a pale brown coat liberally sprinkled with light spots, and a very prominent rump patch. For several days it remains hidden in the grass while the mother grazes nearby and keeps constant vigil. Several times daily she will return to let the calf suckle, but this is done as hurriedly as possible. Many are the predators that are only too anxious to catch the little one, such as mountain lions, wolves, bobcats, coyotes, bears, and even golden eagles. Should the calf be molested it emits a shrill squeal and the cow charges in with sharp hooves flashing. She usually is successful in driving away the smaller predators and sometimes intimidates even the largest with her bristling show of fury. After the calf is large enough to follow the mother, she warns it of danger with a hoarse, coughing bark.

The presence of canine teeth in elk is a peculiarity not found in other American deer. They are of modified form, being bulbous growths without known function. They occur in both sexes but those in bulls have the greatest development. At maturity they become highly polished and stain a light brown.

RODENTS
Including the Lagomorphs
(hares and pikas)

Rodents are the most numerous mammals of the Southwest. This is not an unusual condition; they enjoy numerical superiority over other mammals throughout the world. As a rule rodents are small animals; the largest to be found in the uplands of the Southwest are the beaver and the porcupine. Although these two are considerably larger than all others of the group, they cannot be classed as big animals. Because of the large number of species represented and the varying conditions under which they live, rodents have wide differences in physical characteristics. They can all be identified as belonging to this group, however, by one common characteristic—that of having long, curving incisors. As a rule these number two above and two in the lower jaw, the only exception being the hares and some of their closely allied species. These properly belong to the order *Lagomorpha* but will be included here with rodents.

The incisors are deeply set in the jaws, that part above the gums being a hollow tube filled with pulp. Unlike the incisors of other mammals, they continue a slow steady growth throughout the life of the animal. This is a means of compensating for the wear the cutting edges must undergo. The fronts of these teeth are covered with a heavy coat of enamel, while the back surfaces are either bare dentine or at best covered with very thin enamel. The wear thus results in a bevel-edged surface much like that of a chisel which, with the whetting it receives during the normal movements of eating, remains sharp. A uniform sharpening of both upper and lower incisors is assured by a peculiar arrangement of the hinge of the lower jaw. A more-than-average play in this ball and socket joint allows the lower incisors to slide either behind or in front of the uppers so that both sets receive approximately the same wear on both sides. Should one of the incisors be broken or otherwise damaged so that normal attrition cannot take place, its opposite will grow to such a degree that the animal is unable to take food and then may starve to death. Canine teeth are absent in all rodents, and premolars are lacking in many species. The large gap thus left between the narrow incisors and the comparatively massive molars accounts in part for the wide skull that tapers quickly to the laterally compressed face so typical of rodent features.

Food habits of the various types of rodents differ to a great degree. Perhaps the term omnivorous might be applied to most of them because virtually all rodents will eat insects and meat in addition to the usual fare of vegetable matter. A few might be classed as insectivorous or even carnivorous. Some species store up hoards of food against lean seasons; others eat like gluttons when food is abundant and hibernate through times of want; still others are equipped to spend the whole year in a busy search for something to eat.

Habitats are equally diverse. Some species live below the earth, some on the surface of the ground, at least two species are aquatic, and a few are arboreal. Regardless of where they live, the great majority are home builders. They strive to locate their homes in the most protected places and usually line their nests with soft materials. Outstanding exceptions are the jackrabbit and the porcupine, both of which lead nomadic lives.

In spite of their secretive habits, rodents suffer a tremendous mortality. Practically all carnivorous animals, most predatory birds, and many snakes prey on rodents, and for many of them these persecuted animals form the chief food. This situation is not as harsh as it might seem, for most rodents are prolific to a high degree. Elliott Coues summed up their place in Nature's balance very aptly: "Yet they have one obvious part to play,... that of turning grass into flesh, in order that carnivorous Goths and Vandals may subsist also, and in their turn proclaim, 'All flesh is grass.'"

Snowshoe hare
Lepus americanus (Latin: hare ... of America)

RANGE: Found throughout the greater part of Canada and Alaska with extensive penetrations into the Southwest in Utah, Colorado, New Mexico, and western Nevada. Its occurrence in northern California is rather rare, and is confined to only a few higher mountain ranges.

HABITAT: In the vicinity of streams or in conifer forests in the Canadian and Hudsonian Life Zones.

DESCRIPTION: A small, chunky hare with medium long ears and large hairy hind feet. An average individual will have a total length of about 18 inches with a tail less than 2 inches. Hind foot about 6 inches in length. Summer pelage brown, except feet and belly white, and tail brownish black above. Winter coat white except for the tips of the ears which are black. Young, three to six, born in May or June.

The snowshoe hare, found sparingly in mountains of the Southwest, is the same as that which lives in the muskeg not far from the Arctic Circle. The climate of the mountain zones is surprisingly like that of the north country even though the terrain is different. The closest equivalent is to be found in the brushy borders of mountain streams, and here the "snowshoes" are most often found. During summer they feed on grasses, herbs, and leaves of many different shrubs and the tender tips of young branches. Winter, a period of famine for many animals, is just the opposite for these large-footed hares. Able to run about on the surface of snowdrifts, each new snowfall lifts them closer to the tender twigs that earlier in the year were far above their reach. Clean diagonal cuts much like those made with a knife mark their depredations and, since they are hearty eaters, the whole tops of many favorite food shrubs may be pruned out in one season.

In common with several other hunted creatures and a comparatively few that hunt, the "snowshoe" undergoes a complete change of color between its summer and winter coat. The transformation begins when the first snows are due, and usually the white coat is complete when the snows lie deep on the mountains. It is not, as was once supposed, a case of the brown guard hairs turning white, but a molt. The summer guard hairs are shed and white ones taken their place. The under fur changes color to a less marked degree. Close to the skin the animal is still brown. Outwardly it is pure white except for black ear tips. Marvelous as this protective coloration is, it is not absolute proof against enemies. There are many, and chief among them are lynxes, bobcats, wolves, weasels, and great horned owls. In many places in the far north the snowshoe hare is the chief host of the lynx, their numbers fluctuating in unison.

snowshoe hare

Like most other hares the "snowshoe" spends a great share of its leisure time in a "form." This is usually nothing more than a well concealed hollow. The semi-darkness under low hanging evergreens is much favored by these nocturnal animals for this purpose. They do not, at any time, frequent burrows, the closest approach to this kind of home being in winter when they are sometimes completely snowed under. They suffer but little during severe storms, because their long, fluffy fur is protection against the cold. Their greatest danger lies in the possibility of being buried alive in the event of a freezing rain following the snow.

The young are born in late spring or early summer. They come into the world amid plushy surroundings indeed. The mother has lined the surface nest with soft hair pulled from her own coat, and a softer, more comfortable nursery could hardly be imagined. The little hares are born fully furred, with eyes open, and usually with the incisor teeth already through the gums. Their development is rapid, and long before cold weather arrives they are out on their own.

White-tailed jackrabbit
Lepus townsendi (Latin: hare ... for J. K. Townsend)

RANGE: North of the Canadian border to the southern portion of Colorado and Utah, and from the Cascade Mountains east to the Mississippi River.

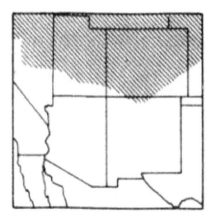

HABITAT: Plains and open country, in the foothills, and even in the high mountains. Found in both Upper Sonoran and Transition Life Zones.

DESCRIPTION: A large hare with a white tail and a lanky build, found usually only in open country. Total length (average) 18 to 24 inches. Tail up to 4 inches. Ears up to 6 inches in length. Weight 5 to 8 pounds. Color varies with the seasons. The summer coat is buffy gray, the winter coat is white. The tail, long for a hare, is white throughout the year. The tips of the ears are black both summer and winter. Young, three to six in a litter, born in May. There may be a second litter during late summer. As with all the hare family, the young are well furred and have their eyes open at birth.

The white-tailed jackrabbit is the largest hare native to the United States. Its great size is further emphasized by its rangy build and long legs and ears. Such physical characteristics are usually marks of an animal that is fiercely pursued by its enemies. This denizen of the open country is no exception. It is preyed upon by innumerable predators, including man, the most relentless and cunning of all. Yet its place in the modern world is still secure, for though it is almost totally lacking in offensive weapons, Nature has given it defensive advantages far beyond most other creatures. Perhaps the most important is the deceptive speed with which it floats across the prairie. Fastest of its tribe and exceeded in this respect by only one native animal, the pronghorn, this lanky jackrabbit simply runs away from most pursuit. Effective though this tactic is, the animal uses it usually as a last resort, preferring to employ the exact opposite, that of crouching motionless in an effort to avoid detection. Absolute immobility is itself an admirable defense, but when augmented by camouflage such as this creature possesses it is even more effective.

Like most members of the hare family, the white-tailed jackrabbit is more active at night than during the day. It spends most of the daylight hours resting in a form that it hollows out under shelter of a low shrub or large tuft of grass. In summer the tawny coat blends well with the color of the surroundings, and the winter coat is possibly even more effective. Then the crouched body resembles nothing more than a mound of snow; the black tips of the ears suggest black weed stems sticking up through the white surface.

white-tailed jackrabbit

Mountain cottontail
Sylvilagus nuttalli (Latin: sylva, wood and Greek: lagos, hare. For Nuttal)

RANGE: Western United States but east of the coastal range of mountains. The northern limits are along the Canadian border; the southern limits in central Arizona and New Mexico.

HABITAT: Mountains of the west through the Transition and Canadian Life Zones. Seldom found below the pines.

DESCRIPTION: The "powder puff" tail is the best field characteristic by which to recognize this rabbit, usually the only cottontail in its range at the elevations given above. It is one of the largest of its kind, averaging 12 to 14 inches in total length with the tail less than 2 inches long. Average weights run from 1½ to 3 pounds. Ears are relatively short and wide for a cottontail. Color varies somewhat with relation to habitat, but in general it is gray with a faint yellowish tinge. Darkest areas are about the back and upper sides; under parts are light to almost white. The winter coat is heavier than the summer, but much the same color. The underside of the tail is the cottony white so well known to city and country dwellers alike. From the scanty records available on the number of young it would seem that three to four constitute the average litter. Perhaps the higher elevations at which they live keep them free from many of the predators to which their lowland cousins succumb, and thus they are able to maintain their numbers with smaller families.

Though often found in the depths of the forest, these shy rabbits prefer to live in the brushy thickets that border high mountain meadows and line the streams. There, in true cottontail fashion, they venture into the open to feed, always ready at the first sign of danger to scurry back to safety under tangled branches. Once fairly entered into the maze of paths that they alone know, there is little danger of capture. There they can count themselves safe from further pursuit by the larger predators and have a distinct advantage over those their own size or smaller. Although so clever at turning and doubling back in their chosen refuges, they seldom use much evasive action when surprised in the open. Their first thought seems to be to reach cover in the straightest possible line, and as a consequence many are snapped up by predators who not only rely on this behavior but often gain the advantage of a surprise attack as well.

Food habits are much the same as those of other cottontails, modified to some extent by the different plant associations with which they are found. In summer, tender grasses and herbs are the favorite fare, but in winter when deep snow isolates them from even the taller herbs, these adaptable animals turn to bark and such small twigs as meet their taste. At this time even the tips of conifer twigs are often eaten. Access to this food, which during the summer is usually out of reach, is facilitated by the growth of long hair on the bottom of the feet, especially on the hind feet. Though these seasonal "snowshoes" do not approach those of the Arctic hare in size, they serve very well to support the lighter cottontails as they move over the soft surface. They are especially useful when the animal stands on its hind feet to reach some inviting bit that would be out of reach in the normal crouching position. During this operation it reaches for food with the mouth alone; the forepaws cannot be used to gather food, but hang loosely in front of the body as an aid to balance.

mountain cottontail

This inability to grasp or handle objects with the front feet is characteristic of all those animals which in the United States we call "rabbits." Though here included with the rodents, the jackrabbits, snowshoe hares, and cottontails all lack the dexterity with the forepaws with which the rest of the group is endowed. The structure of the bones is much like that of the ungulates in that the feet cannot be turned sideways. Thus front legs are used mainly for running, digging, and washing the face and ears, a procedure much like that employed by domestic cats, except that it is carried out with the sides of the paws rather than the insides of the wrists as Tabby does.

Though it lives in a different habitat than other closely related species, the mountain cottontail shares many of their habits. It is a nocturnal animal, seldom seen at large except at dusk or in early morning hours. During the greater part of the day it seeks refuge under some brush pile or deep in the recesses of the slide rock. On occasion it will make itself a form in long grass or under a shrub, but usually prefers more substantial protection. In areas which are being logged, cottontails are quick to take advantage of the shelter offered by huge piles of limbs and debris left by loggers. Later in the season, when the piles are burned, it is not unusual to see as many as three or four cottontails scurry from one pile.

Nests for rearing the young are not of such great concern to these rabbits. Perhaps they instinctively choose places where an enemy would never expect to find them. Many are mere hollows in tall grass or shallow burrows in an accumulation of pine needles. They are lined with soft grasses or needles and hair which the mother pulls from her own body. More hair and grass fibers are cleverly matted together to form a loosely woven blanket which she pulls over the nest when she leaves. It is arranged with such cunning and blends so well with the surrounding that unless one sees the rabbit leave it is only by accident a nest is discovered. The three to five young are born blind and naked, but thrive so well in the warm nest that in about a month they are fully furred and able to leave. At this age they are extremely playful little creatures, often indulging in a game much like tag, although to a human observer it is never quite clear just who is "It."

In this connection it is interesting to note than among the "hunted" mammals the play spirit is usually manifested by running games in which there is little if any physical contact. By contrast, the young of predators indulge in wrestling games featuring use of teeth and claws, often beyond the point where fun ceases and anger begins.

Pika
Ochotona princeps (Mongol name of pika ... Latin: chief)

RANGE: Mountainous areas of the western United States, western Canada, and southern Alaska. Found in the southwestern United States in Utah, Colorado and New Mexico.

HABITAT: Talus slopes of the Hudsonian and Alpine Life Zones.

DESCRIPTION: A small animal bearing some resemblance to a guinea pig; found only among or in the vicinity of rock slides. Total length from 6½ to 8½ inches. No visible tail. Color, gray to brown. Eyes small, ears large and set well back on head. The front legs are short and are exceeded but little by the hind legs. They are all quite concealed by the long hair of the sides. This gives the animal much the appearance of a mechanical toy as it glides smoothly over the rocks. The soles of the feet are covered with hair, the only bare spots on the feet being the pads of the toes. The call is distinctive, the most common being an "eeh" repeated several times. This sound is shrill, but has a falsetto quality as though it were being produced during an inhalation. Young thought to number from three to six.

pika

Far up on the mountainside, above timberline but below the eternal snows, a great field of talus rests uneasily on the massive slopes of bedrock. From a distance it seems merely a smooth gray scar that softens the otherwise abrupt lift of the summit. A closer inspection reveals it as a tumbled mass of variously shaped slabs of stone varying from tiny fragments to huge blocks weighing many tons. Its entire bulk is shot through with chinks and crevices of every conceivable shape and form.

Here and there a wisp of grass or an occasional stunted shrub has found a precarious foothold among the slabs. Other low matlike plants occur in considerable numbers. The only sounds are faint whisperings of wind among the rocks and a distant sighing from the forest below. Suddenly a sharp "eeh-eeh" breaks the silence, then all is quiet again. The shrill sounds are repeated, this time from a different quarter. You look toward the sound but see nothing. Finally, if you are lucky, your eyes focus on a little face somewhat resembling that of a tiny cottontail rabbit, peering at you from the safety of a home among the rocks. It is the pika you see and this rock slide is his castle.

The pika bears a superficial facial resemblance to the rabbit, to which it is most nearly related. This is occasioned no doubt by the long silky whiskers and deeply cleft upper lip, for the eyes are small and the ears, while large, are shaped much differently from those of its larger relative. Its other physical characteristics are entirely unlike those of the rabbit. The chunky body, short legs, and almost total lack of a tail are more like those of the guinea pig to which it is more distantly allied. Several species are known. All are inhabitants of the Northern Hemisphere and all, whether Asiatic, European, or American, are found living in rock slides at or above timberline. In the western United States the pika is known by a variety of other common names of which "coney," "little chief hare," and "rock rabbit" are perhaps the best known.

Living as it does in only one type of habitat, the pika has developed highly specialized habits. The most remarkable is its practice of cutting hay for winter food. At timberline the growing season is short, but the herbs and grasses which this animal eats spring up and mature in a matter of weeks. During this time the pika lives high on the succulent leaves and stems, but during the latter part of the season it carefully harvests enough food to last through the coming winter. None of this hoard is carried directly into the burrow. Instead, it is painstakingly transported to suitable areas which are exposed to the hot sun, and there piled in miniature haycocks and left to cure. No human harvester ever worked harder to gather his crop or laid it up with more care than this tiny husbandman. Fortunately its tastes are not critical; thus, although the individual plants are scattered, the pika is able to select a sufficient store from the considerable number of species represented at this altitude.

In Utah and Colorado the "haying" time arrives with the height of the summer blooming season. At timberline this usually occurs during August. As though realizing that a hard frost would ruin its delightfully fragrant crop, the pika sets furiously to work. After cutting down as much herbage as it can handle at one time, it gathers the mass into an unwieldy bundle and carries it by mouth to one of the sites it has selected as a curing place. Usually these areas have been used the previous season for the same purpose, and a mass of the least edible stems remain to mark their location. Depositing the load on this base, the pika scurries away for another bundle. Long familiarity with routes across the uneven rocks enables it to make its way with never a misstep, even though the load carried may be of such size that vision to the front is completely obscured. Working early and late the pika distributes its harvest among the various piles. As a result, the hay dries out evenly and when cold weather calls a halt to the work each little stack is perfectly cured without a trace of mildew. The truly monumental work to which this little creature goes is shown by as many as a half dozen haycocks, each of which may contain up to a bushel or more of feed.

Comparatively little is known of the pika's life history. What has been recorded has been noted during those periods when it was seen on the surfaces of rock slides. What goes on deep in the labyrinths of its habitat can only be conjectured. It seems reasonable to suppose that in some subterranean cavity the pika has constructed a comfortable nest lined with soft grasses. Certainly it remains active all winter, although buried under many feet of snow, for in the spring its stacks of hay have been largely consumed.

The number of young is thought to range from three to six. They probably are born in early summer, as when they appear on the surface, usually in late July or early August, they are about half grown. Though family ties are closely knit until the young mature, pikas cannot be considered gregarious animals. The scarcity of food alone would be sufficient reason to prohibit large groups in one small area. Each adult takes up squatter's rights on a territory large enough to support it, and thereafter holds it with but little interference from others of its kind.

Few natural enemies prey on the pikas. The very openness of their habitat prevents the larger predators from stealing up unseen. Hawks and eagles account for some, and weasels are able to penetrate their underground maze at will, but the natural fecundity of the species seems to balance these losses very well. To the nature student the pika offers a tempting challenge. It is far from being a rare animal, yet at the same time it is one about which almost nothing is known. As qualifications for learning its secrets, one must be somewhat of a miner and considerable of an arctic explorer.

Tassel-eared squirrel (Abert's)
Sciurus aberti (Latin: shade-tail ... for Col. J. J. Abert)

RANGE: Northern Arizona, northwestern New Mexico, extreme southeastern Utah, and south central Colorado in the United States; also found in the Sierra Madre Mountains of northern Mexico.

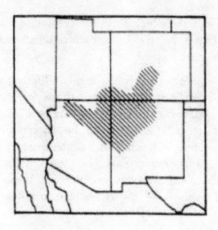

HABITAT: Ponderosa pine forests of the Transition Life Zone.

DESCRIPTION: The only squirrels in the United States that have conspicuous pencils of hair on the tips of the ears. *Sciurus aberti* is a large squirrel with a total length of about 20 inches. Tail about 9 inches. The summer pelage is brown on the back, with gray sides and pure white underparts. The beautiful bushy tail is silvery below and gray above. During summer the long ears have no tassels on the tips. Beautiful as is the summer coat, it is far surpassed by the winter one. Then the heavier fur becomes richer brown on the back, and the contrast between the gray and white areas is further emphasized by appearance of a narrow black band between them. The ears too become more spectacular with the addition of the penciled tufts which give this animal its common name. Breeding habits of this squirrel are variable and evidently depend to a great extent on the food supply. There may be as many as two litters in a fruitful year and none at all in a lean year. The usual number is three or four young to a litter. These are born sometimes in a hollow tree, but more often in a bulky nest of leaves built in a tree top.

No mammal of the United States has a more appropriate generic name than the large tree squirrel. *Sciurus* literally translated means "shade-tail" and refers of course to the beautiful and useful appendage sported by all of our arboreal types of squirrels. It is doubtful if any can equal the striking plume carried by *aberti*; certainly none can surpass it. Its distinctiveness is not occasioned by its size, for several species have tails that are longer. Rather, its elegance is derived from the width, the striking coloration, and the easy grace with which the animal displays its beauty. Whether in full flight across a grassy clearing or in repose on some lofty limb, the first field mark of this unusual squirrel will be the tail; the second, the tasseled ears.

As the map shows, *Sciurus aberti* and its many forms are confined in the United States mainly to the high country along parts of the Colorado River, and also to that great escarpment known as the Mogollon Rim, whose length is divided about equally between New Mexico and Arizona. In this range is found what is often referred to as the "greatest unbroken stand of ponderosa pine to be found in this country." Of the many species of plants and animals found as associates of this forest, perhaps none is more dependent on ponderosa pine than the tassel-eared squirrel. This rough-barked tree furnishes a major source of food and shelter. In return, for Nature always demands that restitution be made, the squirrels plant a part of the seeds that insure continuation of the ponderosas.

It is a common belief that squirrel's diet consists of nuts and little else. This is true only to a degree. A squirrel is fond of nuts and will eat and hoard them during the short season when they are available. For the greater part of the year, when its stores have been depleted, it turns to many other types of food among which are fruit, herbage, leaf buds, and flowers. Favorite food of the tassel-eared squirrel is, of course, the large single-winged seeds found under scales of ponderosa pine cones. Next favored are acorns from the oak that mingles with pine at the lower edge of the Transition Life Zone. If the season is good, great quantities of cones and acorns are buried for future use. These are hidden singly, not in caches, as is the habit of some squirrels. In the event the squirrel does not return for its hidden stores, some of the seeds will sprout eventually and take their part in the slow cycle of growth and decay that is continually going on in the forest.

During months when these favorite foods are scarce, squirrels find the cambium layer of young pine twigs very acceptable. This is the tender layer that lies between the wood and the bark. In the growing season it is especially sweet and nutritious. This was as well known to the Indians as the squirrels, and they too took advantage of the supply during times of famine. The squirrel obtains this food by cutting off the terminal clusters of needles, then severing a denuded portion of the branch, of a size that may be conveniently carried to a favorite eating place. Here the outer bark is deftly removed, the edible portions consumed, and the base wood cast to the ground. Although large numbers of the terminal twigs are taken, the trees seem to suffer no serious damage from this seasonal pruning.

tassel-eared squirrel

In selecting a nesting site the tassel-eared squirrel turns again to its favorite tree, the ponderosa pine. Because few of these healthy giants have knotholes or cavities of a size to accommodate this large species, the nests are usually built in the thick upper growth of branches. Material for their construction consists of small twigs of deciduous trees, cut with the leaves on them. These are cleverly woven together so that as the leaves wither and dry they tend to hold the bulky mass together. Aspen branches frequently are used when available, the large, almost round leaves combining to form a warm wall and at the same time a thatch impervious to all but the most driving rains. Several exits are provided in case an enemy should enter the nest, and the interior is lined with soft fibers. Usually more than one nest is built by each squirrel, so that in an area where they are common the bulky homes are conspicuous not only for their size but by reason of their numbers. With several ports in a storm, so to speak, the squirrels weather the winter very well. During the coldest days they remain snugly curled up in their nests, but on bright, still days they will be seen searching out their hoarded supplies, even though they may have to dig through several inches of snow to get to them. At such time their gruff bark, deep in timbre, may be heard for a considerable distance.

Breeding takes place in early spring, often before the snow is off. The squirrels are fully polygamous, which is one of the reasons this species can almost disappear and then restores its numbers within a season or two. There may be two litters each year, the first arriving as early as May and the second in August or September. As mentioned before, this species is variable and the young may differ in coloration from their parents and from each other. Melanistic individuals are frequent; these should not be confused with the Kaibab squirrel which they resemble superficially. Several subspecific forms are recognized but are not easily identified by the layman.

One's first introduction to this beautiful species is an experience long to be remembered. It was no less interesting to the early naturalists who first penetrated the wild regions where it lived. Their accounts abound with adjectives such as, "handsome," "graceful," etc. Dr. S. W. Woodhouse, who accompanied the Sitgreaves expedition on the exploration of the Zuni and Colorado Rivers, noticed it at once and formally described it as a species in 1852. Since that time it has been introduced into many of the "sky island," mountains that lie south of its original range. It adapts very well to new conditions, seeming to need only a favorable climate and a ponderosa pine forest in which to live. What effect its presence will have on these new surroundings is not yet known. There is always danger that the native plants and animals will suffer from such new competition in an established association. Such introductions should never be made without a study of all the factors involved.

Kaibab squirrel
Sciurus kaibabensis (Latin: shade-tail ... from the Kaibab, a forest in northern Arizona)

RANGE: An area approximately 30 × 70 miles in size in northern Arizona. The southern limit is bounded by the north rim of the Grand Canyon of the Colorado, and much of the range is included within the boundaries of Grand Canyon National Park.

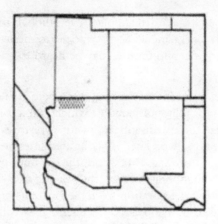

HABITAT: Ponderosa pine forests in Canadian and upper Transition Life Zones.

DESCRIPTION: A tassel-eared squirrel with an *all white* tail. In size this species is the same as *Sciurus aberti* but the coloration is different. The Kaibab squirrel has the same rich, chestnut brown area along the back and upper part of the head, but the sides are deep gray and underparts gray to black. The tail is either all silvery white or it may have barely discernible light gray edging on the upper surface. Nesting and breeding habits are the same as with *aberti*.

Kaibab squirrel

This beautiful squirrel has a distinctive appearance and an uncertain specific rank. It is included here because of all the mammals discussed in this booklet it best exemplifies the effects of isolationism. There is little doubt that the ancestors of both *aberti* and *kaibabensis* were of one common stock. How the progenitors of the Kaibab squirrel came to be marooned on the North Rim of the Grand Canyon is of little moment. Perhaps they were already there when the Colorado plateau was young and the river was just beginning its mighty task. Possibly they emigrated later when the gorge was not as deep as it is now. At any rate, it can be assumed that they have lived on the North Rim for thousands of years, isolated from their cousins on the South Rim by only 20 miles of thin air horizontally, but a trip on foot that involves a descent of a mile through two life zones (Upper Sonoran and Lower Sonoran), a crossing of a wide and turbulent river, and an ascent to the South Rim through the same two desert zones. Surely this is an undertaking for a squirrel of the cool forests that would be too hazardous to be successful, even if attempted.

The factors that have changed this squirrel's coloration are not definitely known, but climatic conditions are probably at least partially responsible. The North Rim is approximately a thousand feet higher than the South Rim and is considerably colder. At this higher elevation much of the Kaibab squirrel's habitat falls within the Canadian Life Zone. This in turn makes certain vegetable food available which is rare or unknown on the South Rim. Thus diet also may have something to do with its unusual appearance.

At various times the Kaibab squirrel has been known as a distinct species, *Sciurus kaibabensis*; at others, it has been considered merely a subspecies of *Sciurus aberti*. The latter is its standing at this time. Regardless of specific rank, it is a form that should be stringently protected. The population is small and goes through the same fluctuations as *Sciurus aberti*. During the summer of 1946 only one individual was known in the area around Grand Canyon Lodge, where they usually were found in some numbers. At such times the heedless destruction of only a few squirrels could conceivably result in the extermination of this rare and beautiful animal.

Arizona gray squirrel
Sciurus arizonensis (Latin: shade-tail ... of Arizona)

RANGE: Central to southeastern Arizona and adjacent parts of western New Mexico in the Upper Sonoran and Transition Life Zones.

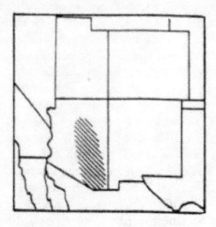

HABITAT: Associated with the native black walnuts of canyons, or often found among the pines on canyon rims.

DESCRIPTION: The common gray tree squirrel to be found in the range given above. The Arizona gray is a large animal. Total length is from 20 to 24 inches with a large tail accounting for from 10 to 12 inches of this measurement. In the typical form the color is dark gray above with underparts and feet pure white. The tail also is dark gray with a silvery white margin. The finest examples of this species may be found along the edge of the Mogollon Rim in Arizona and New Mexico. Farther south the pelage often has a yellowish or brownish tinge. In the mountains along the border the Arizona gray squirrel should not be confused with the Mexican fox squirrel (Apache squirrel) which here barely invades the United States. The Mexican cousin, about the same size as *Sciurus arizonensis*, is definitely yellowish brown and has lighter underparts of the same color. Like other large tree squirrels of the west, the Arizona gray builds a bulky nest of leaves and twigs, usually in the upper branches of a deciduous tree. Young, four or five to a litter; under exceptionally favorable conditions two litters may be reared in one season.

When compared with the royal tribe of Abert's squirrels, this common gray animal of the Southwest seems but a peasant. When it is seen alone comparisons are forgotten. Deliberate in its movements, whether crossing the forest floor or traveling the leafy aisles of the tree tops, it seems always to have calculated its next maneuver. The result is a careless grace that presents the sturdy body and beautiful tail to the best advantage. Calm in temperament and with but little of the suspicious nature that is characteristic of the smaller squirrels, the Arizona gray may easily be tamed in outdoor surroundings and becomes one of the most satisfactory of wild friends. It is not recommended, however, that they be fed from the hand or handled at any time. "Familiarity breeds contempt" is a saying that does not apply to humans alone. A squirrel's bite can be serious as well as painful.

Both Mearns and Bailey, who wrote of this species many years ago, mention it as occurring mostly among the walnut trees of the Upper Sonoran Life Zone. Perhaps during the intervening years the press of civilization has driven them from their chosen habitat into a higher elevation. At any rate, although they still frequent the more isolated valleys, they are now found also in considerable numbers among the pines of the Transition Life Zone. The rough broken country along the Mogollon Rim seems best suited to their requirements, and they are now quite abundant there.

Arizona gray squirrel

Along the border of the Upper Sonoran and Transition Life Zones this adaptable animal finds a wide variety of food. Although the squirrels generally are known as gatherers and storers of nuts, there are many other types of vegetable food that they will take when conditions warrant. The cambium layer of bark and leaf buds of various species of trees are eaten in spring when nut stores have been depleted. Berries, fruit, and even flowers form a considerable part of the diet during the summer. In the fall the ripening crop of pine nuts and walnuts provides not only food for immediate use but stores for the long winter season when, unless enough has been laid by, the unfortunate may starve to death. The gathering period is a time of unremitting labor. From dawn to dusk the squirrels work feverishly carrying nuts to the hiding places they have selected. Sometimes these are in a hollow tree or a nest, but usually the harvest is buried in the humus and debris that collect about the bases of trees.

There are two phases to the work. In the first the squirrel works in the tree cutting off the cones or nuts and letting them fall to the ground. When a considerable number have been thrown down, it descends and carries them away, one at a time. The latter operation is the most dangerous since enemies have an undue advantage over this aerialist when it is on the ground. During the harvest the squirrels plainly show the effects of their work. In gathering pine cones the fur of their forelegs and undersides becomes matted with pitch. The juice of walnut fruits (related to the eastern black walnut, *Juglans nigra*, which the early pioneers used as a source of dye for coloring their hand-loomed cloth) stains their underparts a dirty brown. These marks of their labor remain with them until the summer coat is shed to make way for the heavy winter pelage.

When the generic name *Sciurus* (meaning shade-tail) is mentioned, I am reminded of an Arizona gray squirrel I observed several years ago. During late fall my wife and I were camped near the headwaters of the Hassayampa River in a mixed forest of hardwoods and conifers. Our arrival had interrupted the work of a squirrel which was gathering walnuts in the immediate vicinity, but he soon became accustomed to our presence and renewed activities. Every sunny hour he was busy storing the nuts, many of them at the base of an old pine tree near camp. Shortly thereafter a fall storm set in and lasted for several days. It developed into a pattern of misty drizzle followed by periods of clearing weather when the sun might appear for a few minutes. During sunny intervals the squirrel would appear, but as soon as it became overcast again he would as quickly disappear. Finally we discovered his retreat. When it would threaten more rain he would run up the trunk of the pine to the first branch. Here he would turn his rump to the hole and hunch up into a small furry ball with his long bushy tail laid forward over his back and head and extending down in front of his nose, forming an admirable protection against the few drops that spattered down through the thick foliage overhead.

Squirrels are not the only animals who use their bushy tails for protection against the elements. Many mammals curl up and wrap the tail around themselves for warmth, but only the squirrel tribe has a tail long, wide, and flat enough to be used as a roof. Though the origin of the term *Sciurus* has been lost, it is not too far fetched to suppose that it was suggested by a squirrel's use of its tail as a parasol.

Spruce squirrel, Pine squirrel
(DOUGLAS SQUIRREL, CHICKAREE)
Tamiasciurus hudsonicus fremonti (Greek: tamia, steward and Latin: sciurus, shade-tail ... of the Hudson, named after Fremont)

spruce squirrel

RANGE: Utah, Colorado, Arizona and New Mexico in the Hudsonian and Canadian Life Zones.

HABITAT: Conifer forests, preferably spruce, in the higher mountains.

DESCRIPTION: A small gray squirrel, usually the only squirrel to be found at the elevation at which it lives. Total length 13 to 14 inches. Tail 5 to 6 inches. Two distinct colors of pelage are seasonal. The winter coat is olive gray to rufous gray above with lighter underparts; the summer coat is brownish gray to yellowish gray with almost white belly and feet. A black stripe along the sides is prominent at all seasons. The tail is narrow and noticeably shorter than the body. It is gray beneath, rufous gray above, with black border and a black tip. Little is known of the breeding habits. The four young are born in early summer and by August are usually out foraging with the mother.

Spruce squirrels (distribution shown in accompanying map) include several of the more than two dozen varieties of red squirrels in the United States belonging to the species *hudsonicus*. Combined with several subspecies of the Douglas squirrels, (species *douglasi*, the "chickaree" of the far western mountains), they make up the genus *Tamiasciurus*. This term, a combining form of *Tamias* (the genus of chipmunks) and *Sciurus* (that of squirrels) clearly indicates relationship of the red squirrels to both groups. It is equally apparent in the field where the short narrow tail, the black stripe along the side, and the nervous disposition remind one of the chipmunks, while the arboreal habits, comparatively large size, and coughing bark are distinctively squirrel-like.

The spruce squirrel is seldom, if ever, found below an elevation of 6500 feet, and then only in the shady canyons on the northern exposure of mountains. From this low it will be found up to timberline, or rather just below that point at which the trees are too stunted to offer the required protection. It prefers the dense shade of heavily forested areas, so is rare near the southern limit of its range, and increasingly common in the northern portion.

In common with the rest of its group, this bright-eyed little animal keeps well informed on everything that goes on in the territory it has chosen as its own. Any intruder is thoroughly investigated, then as thoroughly castigated, and driven out if possible. Since these squirrels seem to recognize each other's domain, a trespasser of its own kind usually leaves at the first sign of trouble. With larger animals and humans the attack consists of psychological rather than physical warfare. From a limb at a safe distance above the ground, the doughty warrior chatters and scolds with increasing vehemence as long as a passive interest is displayed by the imagined adversary. At the first threatening movement it disappears in a flash around the opposite side of the tree. Scratching noises and falling flakes of bark, together with noises of peevish defiance, indicates that it is working its way up the trunk. Suddenly it reappears on another limb some distance above the first and the real show begins. Paroxysms of rage, stamping of feet, waving of tail, and streams of invective all are meant to show that one step closer spells trouble. A few squeaks from your pursed lips and this tremendous bluff gives up to curiosity. In a few minutes the erstwhile challenger is back on the first limb trying to make out what this strange creature is about. This amusing procedure can be carried out over and over again, and usually is, just to observe the stuttering rages of which this tiny creature is capable. With more considerate treatment they soon become quite tame, although even then a quick movement will send them helter skelter to the closest tree.

It is well that this squirrel is a quick and tireless worker. The seeds it extracts from the spruce cones are so tiny it takes an enormous number of them to provide that energy. With such a quantity to handle, it is not so careful in storing the crop as some larger squirrels. A comparatively few cones are buried in the soft loam beneath the trees; the rest are stuffed into holes beneath the spreading roots or simply piled in heaps near the base of the trunk. In a year when cones are plentiful there may be a bushel or more in one of these piles. With several such piles within easy reach of the warm nest fastened in the branches of a nearby conifer, the small harvester has prospects of an easy winter ahead.

Only in the most inclement weather are these active animals confined to their nests. They keep tunnels open to their supplies, and each snowfall adds to the security of the caches. All winter long the stockpiles diminish while the snow beneath some favorite perch becomes littered with the scales and discarded centers of the cones. By spring, which comes late at this elevation, the cones are gone and the squirrel returns to its summer diet of leaf buds, seeds, berries, mushrooms, and herbs.

The spruce squirrel is the last of what might be called the true squirrels in this book and, because the group has much in common as regards food, enemies, and relations to mankind, a short summary might be in order.

As has been mentioned, the principal diet of these animals is vegetable. However, all of them, if opportunity offers, will take birds' eggs and young birds. This is not intended in any way as a condemnation of the squirrel tribe. Their inroads on the bird population are what might be termed "natural losses." Nature long ago established a norm in bird reproduction which takes such losses into account.

The enemies of squirrels are legion. From the air, the larger hawks and owls, and even eagles, are ever alert to swoop in on them. On the ground lynx, bobcats, foxes, and coyotes take their toll. In northern Utah and Colorado the marten is one of the most important local controls on the squirrel population. Fast and powerful, the marten is equally at home on the ground or in the trees, and it is a fortunate squirrel that can escape one. The toll taken by all of these predators is high, yet the natural fecundity of the squirrel is so great that the population sometimes gets out of hand and disease has to eliminate the surplus.

In their relationship to man the squirrels are among the most remarkable of our native mammals. It is not ordinarily the purpose of this book to point out the economic importance of our mammals, but the beneficial work carried on the squirrels is too important to pass by. One of the most valuable natural resources that America has is forests. To the arid Southwest the mantles of living green that cover the mountains are invaluable. These are sweeping statements, but they are sober facts.

Squirrels play a considerable part in perpetuating this national heritage. The fact that they do this more or less accidentally merely serves to call attention to the subtle patterns in which all living things move to serve one another. Take their simple mechanics of storing a pine cone, for instance. A hole is dug to a depth of several inches in the soft duff under a shady conifer. The cone is pushed firmly into the bottom of the hole and tamped into place with several vigorous shoves of the nose. Then the hole is carefully filled and smoothed over so that no marauder will discover it. This procedure may be repeated hundreds of times by one individual. If the animal never returns (and the rate of mortality among squirrels is high), the cone can be considered planted. Not only is it planted at the correct depth and in the most suitable material for successful germination and growth, but it is full of plump fertile seeds. Through some instinct the squirrel knows which nuts and cones are healthy and fully developed. If you doubt this, examine some of those they have left on the tree. Invariably they will be infested by insects or "inferior" in some other respect. One of the favorite sources of pine nuts for reforestation projects in the Northern States is the stockpiles of the red squirrel. The scales of the cones are tightly closed when they are taken, but as they open on the drying floor the healthy, fertile nuts prove the unerring judgment of the harvester.

Northern flying squirrel
Glaucomys sabrinus (Greek: glauco, silvery and Greek: mys, mouse)

RANGE: Widely distributed throughout most of our Northern States and Canada. In the section covered by this book, found only in northeastern and south central Utah, with possible occurrence in northwestern Colorado.

HABITAT: Associated with conifer forests of Transition to Alpine Life Zones.

DESCRIPTION: Our only airborne mammal with a long bushy tail. Total length 9¾ to 11½ inches. Tail 4½ to 5½ inches. Characteristic of this species is the fold of skin along each side from the fore to the hind leg. There is considerable color variation in the numerous subspecies of this squirrel. In general the upper parts vary from dark brown to cinnamon brown. Sides of face gray; underparts white to pinkish cinnamon beneath. Hind feet are brown, fore feet gray. The flying membrane is brownish black above, white to cinnamon beneath. The eyes are large and dark brown. Young, two to six in a litter, born in spring; a second litter is sometimes produced in early autumn.

Because flying squirrels are almost entirely nocturnal, they are seldom seen. This is unfortunate, for they are among the most interesting forest creatures. Probably more people have seen flying squirrels through the predations of a house cat than in any other way. Gentle and unafraid, the squirrels fall easy prey to this night prowler, which sometimes brings them home to show its owners. Strangely enough, the victims often are not injured seriously, and if taken from the cat and allowed to recover from their initial fright they will glide about the room with much of the grace they display in the wild.

Properly speaking, these squirrels do not fly; that is to say, they are incapable of sustaining level or ascending flight. Rather they climb to some height in a tree then launch out and glide to a lower point, usually the trunk of another tree. As the angle is usually quite sharp they attain considerable speed. They check this momentum by inclining upwards just before reaching their objective. This results in a four-point landing against the tree trunk, sometimes with an impact that can be heard for some little distance on a quiet night. During these flights, which may extend 50 yards or more, they are able to change direction or maneuver against wind currents. This is done by manipulating the flying membrane and using the tail as a rudder. After a flight they usually ascend to the safety of the foliage above. They cannot be considered awkward on the ground, but it is not their chosen habitat. Flying squirrels are more arboreal than any of our mammals, excepting a few species of bats.

northern flying squirrel

Little is known of the habits of this unusual squirrel, but they differ considerably from those of its relatives who are active during the sunny hours. Instead of living in a bulky nest hung in tree branches, this nocturnal aerialist chooses a hollow tree or an abandoned woodpecker's hole where the sun's rays never penetrate. Nests have been found also under slabs of bark hanging to old lightning-blasted snags. Lined with soft fibers and shredded bark, they often shelter whole families of flying squirrels for, unlike the other squirrels, these gentle creatures get along together. In fact, they might almost be considered gregarious. Contrary to ordinary squirrel behavior, they never bark or scold. Their only utterance is a fine whistling squeak, and this is heard usually only in the nest.

Though delicate in appearance the flying squirrel is extremely hardy. It is abroad throughout the winter, being confined to its nest only during stormy weather. It stores food for the winter, but its caches are usually above ground in hollow trees and crevices rather than buried in the loam. Their food consists mostly of pine nuts, seeds, and acorns, but they are also fond of meat. Many a flying squirrel has met its death trying to take the bait from a trap set for larger game. This taste is unexplained; it is not known to prey on other animals.

Western chipmunks
Genus *Eutamias* (Greek: eu, well or good and tamias, steward)

There are at least four species of chipmunks native to the area covered by this book. Ordinarily but one, or perhaps two, species of a genus have been chosen for discussion. In this case, however, the chipmunks are such provocative little creatures and their presence causes so much interest that all four species will be included, although briefly. Since the ranges and life zones of some of them overlap in many areas, positive identification of a species will be difficult in those places, but in others one species will be dominant or alone. Here the more subtle characteristics and behavior of that type can be fixed in mind, and in time it will be less difficult to separate one from the other. Remember that most of these species have several subspecies. These generally occur along the upper or lower edges of the life zone frequented by the type. In the field they are usually indistinguishable from the type to any but the most practiced observer.

1. Colorado chipmunk (*Eutamias quadrivittatus*)

Colorado chipmunk

RANGE: Northern Arizona, northern New Mexico, most of Utah, and all but the most northern portion of Colorado. This chipmunk lives largely in the Transition Life Zone. The closely related species *umbrinus*, commonly called "Uinta chipmunk" inhabits the Canadian and Hudsonian Life Zones in the Uinta and Wasatch Mountains of northeastern Utah.

Colorado

Uinta chipmunk

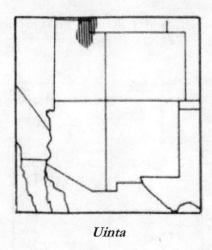

Uinta

2. Gray-necked chipmunk (*Eutamias cinereicollis*)

gray-necked chipmunk

RANGE: Central Arizona eastward into southwestern and south central New Mexico. Total length 7½ to 10 inches. Tail 3½ to 4½ inches. Transition Life Zone and above. *Neck and shoulders gray.*

Gray-necked, Cliff

3. Least chipmunk (*Eutamias minimus*)

least chipmunk

RANGE: Western Colorado, western Utah, northern and eastern Arizona, northern and central New Mexico. Inhabits all zones from Upper Sonoran to Alpine. Total length 6⅔ to 9 inches. Tail 3 to 4½ inches. *The smallest chipmunk with proportionally the longest tail. Tail carried straight up when running.*

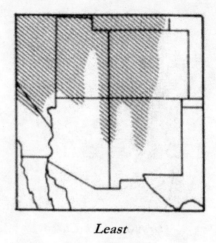

Least

4. Cliff chipmunk (*Eutamias dorsalis*)

cliff chipmunk

RANGE: North and western Utah extending through southeastern Arizona and western New Mexico. Found mainly in the Upper Sonoran Zone. Total length 8⅘ to 9½ inches. Tail 3⅘ to 4½ inches. *The most indistinctly striped of any of these chipmunks.*

Generally speaking, chipmunks are the link between ground squirrels and tree squirrels. Physically they have characteristics of both groups, a combination that is pleasing indeed. A field mark that is a positive identification of the chipmunk group is the striped face. In addition to facial stripes, chipmunks also are striped along the back. The pattern consists of a dark to black median line bordered by two more similar lines of varying intensity along each side. These fine lines are separated by broader bands of contrasting color ranging from chestnut to white. The latter characteristic is shared by several of the ground squirrels, which often are confused with chipmunks. Predominant colors of southwestern chipmunks run to rufous, chestnut, and grayish white with the dark to black lines mentioned above. Underparts are always considerably lighter than the back. Chipmunks' tails are usually shorter than their bodies, flattened horizontally, and short haired when compared with tree squirrels. All species have cheek pouches of considerable capacity.

As will be seen from ranges given above, habitat of the chipmunks encompasses the whole area from sagebrush-covered foothills to timberline. Their densest population, however, is to be found in thick forest about midway between these two extremes. Here their bright colors and sprightly actions do much to enliven somber surroundings. Despite their wonderful climbing ability, they are most often seen at ground level or just a little higher. They are fond of areas containing fallen trees. The prostrate trunks serve admirably as highways for their forays in search of food, and under the litter which accumulates around them are many havens into which a hard pressed chipmunk may pop when pursued by an enemy. The territory appropriated by each of these little creatures is explored with the most minute care, and all places of refuge are noted for future emergencies. Any attempt to chase them will reveal their uncanny memory for these temporary hiding places and that they are seldom at any great distance from one.

Their permanent homes usually are underground, excavated beneath the roots of trees or in rocky terrain. At the end of a narrow tunnel a room of considerable size is worked out. The dirt is often carried out by a side tunnel, which is permanently plugged with soil when the excavation is completed. The underground chamber is lined with soft grasses and fibers as insulation against the cold. At the higher elevations the ground may freeze to a depth of several feet during the long winters. Permanent nests are sometimes built in hollow logs, but almost never in holes in upright trees. Chipmunks have little taste for upstairs apartments. In addition to the large cavity which contains the nest, several storage chambers are constructed to hold the winter's food. These may be connected to the main apartment by tunnels or may be entirely independent of living quarters and some distance away. As a special feature, many of the more elaborate homes have a separate chamber reserved for sanitary purposes. Like most of our native rodents, chipmunks are fastidiously clean in their habits.

It is difficult to discuss the habits of a group as large and of such wide distribution as our southwestern chipmunks in any but a most superficial manner. In general they are much more terrestrial than squirrels and prefer brushy, rock terrain to the more open forests frequented by their larger relatives. Nevertheless, they are adept climbers and do not hesitate to take to the trees in search of food or to escape their enemies. These arboreal excursions are usually limited to one tree; they do not ordinarily attempt the daring leaps from one to another that are characteristic of the squirrels. They progress quietly while on the ground, threading their way through the undergrowth so expertly that their presence is often undetected.

Normally chipmunks are shy creatures at first acquaintance, but if their friendship is encouraged they often become bold to the point of being unwelcome. Woe to the camper whose grub box is invaded during his absence. These tiny opportunists can carry away a surprising amount of food in a very short while. Their natural diet differs widely according to habitat. Chipmunks of the foothills eat a great variety of grass seeds, berries, and cactus fruits. These are possibly the favorite foods of the whole group, but as the elevation increases this supply becomes limited and is supplemented by juniper berries, acorns, and pine nuts. Considerable quantities of these less perishable foods are laid away for future needs. During the summer months herbage, fungi, small tubers, and some insects add variety to an otherwise dry menu.

It is doubtful if any southwestern chipmunks enter true hibernation during the winter. Those of lower elevations are active throughout the colder months, except when a period of exceptionally inclement weather will force them to remain underground for a few days. At higher elevations they will disappear, perhaps for weeks at a time, but it is assumed they remain active in their underground quarters. The fact that during the fall they do not lay on a coat of fat, like many species which are known to hibernate, substantiates this theory.

Breeding habits of chipmunks are not too well known. The number of young averages from four to six. Those species living at low elevation sometimes bear two litters each year; those at higher elevations are limited to one. Like the ground squirrels, the young are able to leave the burrow when but little more than half grown. At this early age they present a rather ludicrous appearance with their large heads and sparsely-haired tails. This is a time of great danger, for the youngsters are easily caught by predators which would be eluded with little difficulty by a mature individual. Principal predators of the chipmunks are bobcats, hawks, foxes, and coyotes. The last two often dig out the burrows. The marten is possibly their worst enemy, but fortunately for the chipmunk tribe is a rare animal throughout its range.

Chipmunks are quite common in several of our southwestern National Parks and Monuments. Despite signs to the contrary, the public cannot resist feeding these little beggars, and many are the situations that develop from this practice. I recall camping at Bryce Canyon National Park where the least chipmunk is a common resident. Upon our return from Rainbow Point one day we spied a chipmunk with bulging cheek pouches leaving our tent for its den somewhere on the edge of the canyon rim. We found that our visitor had entered the grub box and gnawed a neat hole in the top of a carton of rice. Although we had been gone but a short time, more than half the contents had already been carried away. This was a state of affairs that needed mending so we decided to teach the marauder a lesson. On his return trip we waited until he had entered the carton and then clapped a dishtowel over the hole. The cellophane window in the side of the carton gave us an excellent view of our prisoner. Interrupted in his pilfering, he at first tried to get out of the carton but, finding no exit, returned to stuffing his cheek pouches with more rice. When they were filled to capacity he calmly sat back and returned stare for stare. In the end we let him go and gave him the rest of the rice, exacting such payment as we could by taking pictures of his labors.

Golden-mantled ground squirrel
Citellus lateralis (Latin: citellus, swift, and lateralis, belonging to the side, referring to the stripe along the side)

RANGE: Western United States and Canada. In the area covered by this book to be found in western Colorado, from northeastern Utah south through central Utah to central Arizona thence east into western New Mexico.

HABITAT: Higher mountains of this area. Usually found in evergreen forests of the Transition, Hudsonian, and Canadian Life Zones. It sometimes occurs near the upper limits of the Upper Sonoran Zone.

DESCRIPTION: A chipmunk-like ground squirrel lacking the stripes along the sides of the face characteristic of the chipmunks. Total length 8½ to 12½ inches. Tail 2½ to 4½ inches. There is much color variation in this species. Head coppery to chestnut, upper surfaces of body brownish gray to buffy. A light to white stripe bordered with black is present on each side of the back. Under surface of tail gray to yellow. Tail short but fully haired. Under surfaces of body lighter, gray to buffy gray. Legs short, body chunky in comparison with chipmunks. Young, four to eight, with but one litter each year.

The golden-mantled ground squirrel has been chosen from the rather large group of southwestern ground squirrels because it is most typically a mountain dwelling species. As such it does not have the advantages of a long summer season like its lowland relatives. This results in two definite periods each year. One is feverish activity during summer, a time of breeding, rearing the young, storing food, and laying on fat for the cold months ahead. The other in winter is the exact opposite—a long interval of hibernation when, buried deep under the snow in a snug burrow, the squirrels sleep away the winter.

Though hampered by the short summers of higher elevations, the golden-mantled ground squirrels manage to lengthen the season slightly by a very simple expedient. Instinct prompts them to dig their burrows on a southern exposure, often under the base of a log or in a rock slide. Here the snow melts away first and they often have a bare spot of ground in front of the burrow several weeks in advance of the season. The squirrels emerge from their long sleep weak and emaciated, and their first days above ground are spent soaking up the warm sunshine and waking up, so to speak. During this period they live on stores laid away the previous summer, and by the time the snow has melted they are fully active and ready for mating.

golden-mantled ground squirrel

As with the ground squirrels of lower areas, the summer diet consists largely of whatever starts to grow first. During late spring, grass, buds, young leaves, and flowers are eaten. Later, seeds of the annuals are gathered, berries are taken whenever possible, and insects often form a considerable part of the diet. As fall comes on, acorns, pine nuts, and a great number of smaller seeds and fruits become available. At this time the ground squirrels must not only lay on enough fat to maintain themselves through hibernation, but must also store away enough food to tide them over between the time of their emergence and the appearance of new growth. Evidently this is an adaptation forced upon them by the exceptionally long winter season. Most rodents which lay on coats of fat preparatory to hibernation depend almost entirely on it to carry them through. With a hibernating period of from 5 to 7 months, however, it is not difficult to realize the problems this ground squirrel must face.

Though the golden-mantled ground squirrel resembles the chipmunks in appearance, its temperament is quite different. Chipmunks are bright, nervous little sprites, always pursuing their activities with explosive energy. The ground squirrels move more sedately, as though they had planned every move and there was no hurry. They love to lie in the sun in some exposed place and watch the rest of the world go by. In habitat, too, the species differ materially. Chipmunks choose thick undergrowth where they can go about their business unobserved. Ground squirrels prefer more exposed locations where they take their chances in the open, but with one eye always cocked aloft as insurance against attack by hawk or eagle. Creatures of the earth, they are always reluctant to climb. Rarely do they ascend more than a few feet, and then only to reach some especially toothsome delicacy that their keen noses have detected in a low shrub or small tree.

With its wide distribution, visitors to the southwestern mountains can hardly fail to notice this golden-headed member of the ground squirrel family. It is easily tamed; too easily in fact for, like the chipmunk, it can quickly wear out its welcome. In many of the National Parks and Monuments they compete with chipmunks for the crumbs around camp sites and picnic tables. Visitors find their cunning way irresistible and feed them despite warnings to the contrary. Because they do tame so easily there is always danger that some well-meaning person will attempt to pick them up. This can lead to unpleasant results. Their long sharp incisors can inflict a serious wound.

One of the most fascinating places to observe both chipmunks and these ground squirrels is from the windows of the long tunnel leading northward out of Zion National Park. On the talus slopes beneath the windows a great number of these rodents take up summer quarters, depending for food on the largesse distributed by visitors as they eat their picnic lunches on the broad ledges of the windows. Their constant movements as they run among the rocks seeking stray crumbs result in many a collision and often an angry dispute as well. This proves a dangerous game, as rocks sometimes will be loosened by their movements and roll down the steep incline. I recall seeing a ground squirrel crushed by one of these miniature rock slides in 1946.

White-tailed prairie dog
Cynomys gunnisoni (Greek: kun, a dog and mys, mouse ... for Captain Gunnison whose expedition took the type)

white-tailed prairie dog

RANGE: Western Colorado and eastern Utah to central Arizona and New Mexico.

HABITAT: Grassy meadows and mountain parks mainly in the Transition Life Zone although they are often found both above and below this area.

DESCRIPTION: A ground-dwelling rodent somewhat resembling a ground squirrel but several times larger than the biggest species of that genus. Total length 12½ to 15 inches. Tail 2¼ to 2½ inches. Weight 1½ to 2½ pounds. Color buff to cinnamon buff, the short fully-haired tail tipped with white. Sides of face darker with a dark area over the eyes. Legs, feet, and underparts pale cinnamon buff. Young, usually five in number, born in early summer.

Cynomys gunnisoni is the representative species of the western group of prairie dogs. The two remaining of the group, *Cynomys leucurus* and *Cynomys parvidens*, both white-tailed species, are very similar and possibly will be classified with *Cynomys gunnisoni* in the future. *Cynomys leucurus* is found in northwestern Colorado and northeastern Utah, while *Cynomys parvidens* is native to mountainous valleys in central Utah.

The common name "white-tailed prairie dog" is usually applied to *Cynomys gunnisoni*, the most widely distributed member of the race. The range of this species borders on but seldom overlaps that of the black-tailed prairie dog which lives farther east and at lower elevation. Climatic and geographic barriers separating these two races are largely responsible for pronounced differences in their habits. Prairie dogs are gregarious creatures, perhaps more so than any other rodent. Formerly the black-tail species inhabited countless thousands of acres in the Great Plains region. A single colony might occupy an area several miles in diameter and number many thousands. On this relatively flat land, every home site was equally advantageous and the grass and herbage all ideally suited to the prairie dog's use. Periodic flooding of their burrows on these level prairies was avoided by building conical mounds with a rim of earth around the entrance. This ingenious practice, simple though it seems, represents a long step in the adaptation of these animals to their environment.

White-tailed prairie dogs, on the other hand, are limited to the narrow valleys and infrequent open meadows of the mountains. Here there is neither room nor food to maintain the huge colonies characteristic of the black-tailed. Under these conditions the number of individuals in a town will vary from a few to 200, seldom more. If the town becomes crowded, many of the inhabitants may migrate to some more favorable location. This sometimes entails a trip of several miles, a hazardous undertaking for a small animal whose only escape from large predators is in an underground burrow.

Food of this mountain prairie dog is varied. The standard diet of grass and roots is augmented with browse, bark, and tubers. Bulbs of mariposas are taken wherever available. Coarse-leaved annuals such as sunflowers are not passed by. In addition to this vegetable diet, worms, beetles, and larvae as well as mature forms of most insects are eaten whenever possible.

Burrows of white-tailed prairie dogs, though comfortable, are not made with the painstaking care found in those of the lowland species. There is no need for a conical mound or built-up rim because there is virtually no drainage problem on the sloping terrain of the mountains. Naturally the burrows will not be excavated in the path of flood waters, but on higher ground. Earth brought out from the underground workings is piled to one side or in front of the entrance. The mound thus formed is used as a place to sun bathe or, even more important, as a look-out post from which to see all that goes on. Because these small colonies do not have the advantage of numbers, each individual must be especially alert to approaching danger. Burrows often have more than one entrance, each with its well-packed sentry post at hand, the underground plan is simple. It consists of a more or less vertical shaft from which one or more tunnels extend horizontally. It is common supposition that the prairie dog digs deeply enough to strike water. This is not so; many burrows do not go deeper than 6 feet. In any event, they penetrate just far enough to insure a comfortable average temperature in both summer and winter. Water requirements of prairie dogs are met largely by the succulent nature of their food. It is also presumed that during late summer months when the diet consists to some extent of seeds, a chemical process within the system transforms some of the starches to water.

The nest is usually situated in an underground room dug at the end of a tunnel, less often somewhere along its length. It is a bulky structure, built of shredded bark or coarse grasses and lined with the softest fibers obtainable. In these modern days prairie dogs do not object to paper, rags, and wool.

The life of the prairie dog is simple. Early in the spring it emerges from hibernation, a bit groggy but still well padded with fat. This nourishment sustains it until the first green shoots of grass appear. From then on food is obtainable in an ever increasing supply, limited only by the distance to which these indifferent runners dare venture from their burrows. Summer is a time of eating, of dozing on the mounds in the warm sun, and of conversing with neighbors in the shrill barking whistles characteristic of this group. It is also a time of constant vigilance against predators, of dust bathing to rid themselves of mites and fleas, and of rearing the young. The four to six young are born in late spring and first appear at the burrow entrance when about the size of an average adult ground squirrel. Within a few days they are foraging for themselves, and about 3 weeks later are able to make their own way. At this time the mother frequently deserts them and builds herself a new burrow, leaving her offspring to divide the old homestead as best they can. As fall draws near, a thick coat of fat is put on, and by the middle of October most of the town's inhabitants have retired for the long winter's sleep.

Yellow-bellied marmot (woodchuck)
Marmota flaviventris **(Marmota, Dutch name of European species of woodchuck. Latin: flavus, yellow, and venter, belly)**

RANGE: Northwestern United States. Common in northern to south central Utah, northern and southeastern Colorado, and extreme north central New Mexico.

HABITAT: Canadian, Hudsonian, and Alpine Life Zones in rock slides, rocky hillsides, under rock piles, and around outcroppings in mountain meadows. Seldom found below the Canadian Zone but often occurring in the Alpine Zone to the very summits of the mountains.

DESCRIPTION: A large, dark, brown marmot with a comparatively long bushy tail. Total length 19 to 28 inches. Tail 4½ to 9 inches. Body color, yellowish brown to dark brown above; under parts yellow. The body fur has a grizzled appearance. Sides of neck buffy, and sides of face dark brown to black. Light brown to white between the eyes. The feet are buff to dark brown. Tail dark brown above, lighter below. Young, five to eight, born in early summer.

This large western marmot is not too far removed from the ground squirrels in either relationship or habits. It is the largest ground-dwelling rodent native to the Southwest. As mentioned above, marmots occupy a tremendous altitudinal range, reaching from above timberline down into the Transition Life Zone. This distribution from arctic to almost desert conditions is responsible for many variations in their habits. Most important is the practice of estivation by those individuals which live at the lower elevations. This summer sleep is used as a defense against that period of drought between rainy seasons. It usually starts early in June and ends about the latter part of July. In the higher life zones there is no lack of green food throughout the summer, consequently marmots there remain active.

Because of large size and ability to make good use of its sharp teeth and claws, the marmot's life is not so restricted as that of many smaller ground-dwelling rodents. It has enemies, to be sure. Bears, mountain lions, wolves, lynxes, wolverines, and eagles all are alert for a possible catch. Yet it is so well on guard and has so many burrows that it is next to impossible to catch one above ground. Should the marmot be surprised away from a burrow, its bold show of defense often gains enough time to work its way to a place of safety. When cornered its appearance alone is enough to make the average predator pause and consider. With hair standing on end and long claws at the ready, the marmot clatters its sharp teeth and hisses loudly at the enemy. This pose is not all bluff. These big rodents are courageous and able adversaries against any animal up to several times their size. As far as man is concerned, they are timid and secretive. On many an occasion their loud, full-toned whistles will be heard, but the whistler will be nowhere in sight. If cornered, however, they will put up the same courageous defense they display against other enemies, and certainly are not animals with which to trifle.

Burrows are usually in open places where a good view of the surroundings is obtained. Too, they are almost always in clefts of rocks, under boulders, or in coarse rocky soil. This lessens the probability of their being dug out by some large predator. Each marmot usually will have several burrows, some being "escape" means and one a permanent home. Well-worn trails lead from one to another, for these are active animals which travel extensively within the limits of their territories. Escape burrows may be deep or shallow, as circumstances dictate, but the home burrow generally is a labyrinth of long passages that terminate in a nest chamber up to 2 feet across. Several auxiliary tunnels are usually reserved for sanitary purposes. None is used for food storage; records indicate that this creature does not lay up stores for later use. The nest is the usual bulky affair, built of coarse materials and lined with the softest grasses and fibers obtainable.

Late to bed and early to rise is characteristic of the marmots. Classed as a diurnal animal, they nevertheless travel about a good deal at dusk. During the breeding season they may even make an extended trip at night to find a mate. Sunrise signals the beginning of the marmot's day. The slanting rays have no more than touched the boulder above its burrow before the inmate will climb up to take advantage of their warmth. It may stay atop its vantage point for an hour or more. There are many things a marmot can attend to while taking the early morning sunbath. A leisurely toilette, whistled comments to neighbors, a long scrutiny of the terrain for possible danger—all these are matters requiring thorough attention.

yellow-bellied marmot

Should this procedure be interrupted by a prowling enemy, excitement runs high. If the intruder is still some distance away, the marmot often will stand up on its hind legs, picket pin fashion. Each explosive whistle will be accompanied by several flicks of the tail. When it is judged time to retire it will dash for its burrow, making sharp chirps as it goes. Once inside the burrow it may chance another look outside, and if the caller looks menacing enough the burrow entrance will be plugged with earth from inside, the chirps becoming fainter as the barricade is forced into place. Emergence from the burrow after a fright of this kind is governed to some extent by the time of year. If it is autumn and the marmot is about ready to hibernate, it may go to sleep in its cozy nest and not reappear until the next day. Even in spring and summer it will remain underground for a considerable time before venturing out again.

The marmot is by nature a stocky animal. Short-legged and barrel-bodied, it can lay on a surprising amount of fat for the period of hibernation. Length of this winter sleep depends on the elevation at which the animal lives. On the higher mountain tops it begins about October 1. At lower elevations it may be considerably later. Older individuals usually go into hibernation first, presumably because they are able to lay on the necessary fat sooner than younger ones. As a rule they retire by stages, disappearing for several days at a time; their movements are lethargic and they act as if already half asleep. The young of the year have spent the greater part of the summer growing up, and it is rather a grim race with time to determine whether they will be able to put on enough fat to carry them through the long winter with a reserve supply, or whether they can survive the cold weather that greets them. Especially at the higher elevations, they do not retire until forced to do so by cold weather.

Hibernation is as profound with these big rodents as with many of the ground squirrels. They will curl up into furry balls in their cozy nests, noses covered with fluffy tails, and sink into a deep sleep that approaches suspended animation. Bodily functions slow to a fraction of the normal rate, and the system draws on its store of fat to survive. The drain on this nourishment is slow, as it necessarily must be, for this single source of food must last for a period of perhaps 5 months.

The date of emergence varies. Although February 2nd is recognized as groundhog day on our calendar, this date would be chilly indeed on the peaks of our Southwest mountains. Nevertheless, the marmots do appear before the snow is entirely gone, and once their sleep has ended they rarely resume it, whether or not they see their shadows.

Breeding takes place shortly after emergence. The young are born in April or May. They are born blind; the eyes do not open until about a month after birth. The youngsters develop rapidly, and by the time they are half grown a daily session of sunbathing and playful tussles outside the entrance of the den is part of their routine. By September they are fully grown, and at this time they usually strike out for themselves, although cases have been recorded in which the family remained together through the first winter's hibernation.

Marmots have always been favorites of this writer. Their clear-toned whistle is as much a symbol of the rugged peaks and lovely fir-rimmed mountain meadows as the coyote's barks are of the desert. Several writers characterize marmots as "stupid." Surely this is an unfortunate choice of word. Stupid by what standards? Can one species be compared with another when all must live under the different conditions to which they have adapted themselves? The mere fact that a balance of Nature has been attained indicates that each has the adaptations, the habits, and the degree of intelligence necessary for that species to live in harmony with the whole.

Deermouse (white-footed mouse)
The genus *Peromyscus* (Greek: pera, pouch, and muscus, diminutive of mys, mouse)

deermouse

RANGE: All life zones throughout North America.

HABITAT: Some species of deermouse can be found in almost any association imaginable.

DESCRIPTION: A large-eared mouse with white feet. Since there are many species in this genus and most of them are quite similar, characteristics common to the greatest number will be given. Bear in mind that these may not hold true with every species of the genus.

Deermice are rather small, averaging 7 to 8 inches long. Tail 3 to 4 inches. Most species are a buffy gray above shading to brighter buff on the sides and light buff to white beneath. Feet are always white. The ears are large for a mouse, usually sparsely covered with short, fine hairs, but in some species almost naked. Eyes appear black but have a brownish shade when viewed closely in a good light. Tail long, up to the length of head and body, as a rule sparsely haired; bicolor in some species. Young, four to six, born almost any time of the year, with several litters except at higher elevations where only one litter may be born, and this during late spring.

In the Southwest the mild climate and plentiful food supply of the lower life zones combine to attract a great number of small rodents. By far the greater number of species is found in the Upper and Lower Sonoran Zones. This does not mean that mice are rare in the high mountains. They live there in great numbers, but of fewer species. One is the long-tailed deermouse (*Peromyscus maniculatus*), probably the most outstanding member of the genus, and the most widely distributed mouse in the United States. As might be expected, it is quite variable in appearance, having at least three distinct color phases. These vary from golden tan to a dark gray. All phases have a sharper bicolor tail, white beneath and like the rest of the upper body on top.

The deermouse is well known to those who are fortunate enough to own summer cabins in the mountains. This is the little rodent which moves into the cabin as soon as the vacationer departs. Fortunately it is not so destructive as the common house mouse (which, by the way, is an introduced species) and limits its destructiveness for the most part to building a large and comfortable nest in which to live during the winter months. Deermice do not hibernate, so they must prepare against the bitter cold. However, it is not their habit to store food either, and doubtless many of them starve to death over a hard winter.

Mountain vole
Microtus montanus (Latin: small ear ... of the mountains)

mountain vole

RANGE: The mountainous regions of northwestern United States extending eastward to central Colorado and southward below the northern borders of Arizona and New Mexico.

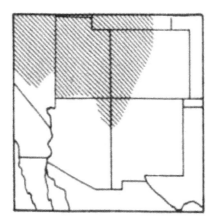

HABITAT: Valleys and grassy meadows seldom lower than the Transition Zone.

DESCRIPTION: A small sturdy rodent with short tail, total length 5½ to 7½ inches. Tail 1½ to 2½ inches. This is a very short tail for a rodent of this size, amounting to only about a fourth of the total length. Color, grayish brown to black above; underparts lighter to a silvery gray. This is but one of many species found in southwestern mountains. The Mexican vole and the long-tailed vole are two which share its range. They are quite similar in appearance and their life histories also are much the same.

In several ways this heavy-set rodent resembles the pocket gopher. The small ears and eyes as well as the short tail are all reminiscent of that animal. Like many other rodents, voles are quite prolific. From four to eight young are born in a litter. The number of litters each year depends to a great extent on the altitude. They have been recorded in the Canadian Zone, where the summers are too short to permit the rearing of more than one litter. In the Transition Life Zone they commonly bear two litters and sometimes more each year.

These are the small rodents which most people call "field" or "meadow" mice. In the prairie states this genus is well known for its habit of congregating under shocks of small grain and corn. Here they build their nests and temporarily live in peace and plenty. When the shocks are taken from the field, they are rudely evicted from their snug shelters to fall prey to the farmer's dog or to face the prospect of building a new home before winter descends upon them. In the West, too, this "field mouse" makes itself at home in agricultural areas, but its native haunts are the natural meadows in mountain valleys. Here they build tunnels in the tangled growth of grass, and excavate shallow burrows in the soft earth. Marshy places are particularly to their liking, because they are quite at home in water. Too, the thick cover in these areas gives them considerable protection from their many enemies. A normally high reproduction rate (several litters per year with up to eight young in each litter) coupled with a secretive way of life insures their perpetuation. In cases where a natural balance has been upset, their population can soar to fantastic heights. In one agricultural district in Nevada a survey revealed an estimated 8,000 to 12,000 "field mice" per acre.

Voles do not hibernate. They are active night and day, summer and winter. During winter storms they may remain in their snug nests for a few days at a time, but with the return of clear weather, openings to their tunnels will soon appear in freshly fallen snow.

Western jumping mouse
Zapus princeps (Greek: za, intensive and pous, foot. Latin: princeps, chief)

RANGE: Western United States from central Arizona and New Mexico to Alaska.

HABITAT: High mountains in dry places with abundant low ground cover.

DESCRIPTION: A small rodent, two-toned in color, that leaps through the grass much like a kangaroo rat. Total length 8 to 10 inches. Tail 4½ to 6 inches. Color buffy along sides, shading to almost black on the back and white on the underparts and feet. Tail bi-color, dark above and light gray beneath. Ears relatively long, dark in color with light buffy marginal lines. Eyes beady, set in long face with sharp nose. Front legs short but hind legs and feet large and muscular. Young, four to six in a litter, with no more than one litter a year in the higher elevations.

The jumping mice are among the most specialized small rodents in the United States. The genus is typically North American, only one species being found outside this continent. At some time in the distant past this little creature adapted itself to a mode of flight much like that of the kangaroo and jerboa. In this respect it exceeds the kangaroo rats and pocket mice of the United States, species to which it is distantly related. Its general build is distinctly like that of the kangaroo, with the same delicately formed front quarters and heavier hind quarters. The tail, though not club-shaped like the kangaroo's, is long enough to serve the same purpose—that of a rudder to guide the direction of flight. The hindlegs are muscular enough to propel the body on proportionally longer jumps than even the kangaroo. Here the resemblance ceases, however, for the jumping mouse is not related, even distantly, to this marsupial. The only pouches the jumping mice have are internal cheek pouches used exclusively for transportation of food.

Jumping mice have one more peculiarity that set them apart from most other North American mice; they hibernate. The period of hibernation is not a short one at the elevations at which these mice live. It may last for as long as 6 months. Preparation for this extensive period of inactivity consists mainly in gathering and eating grass seeds until a thick layer of fat is stored under the skin. With the first cold weather the jumping mice retire to previously prepared underground burrows and sleep the winter away.

Since they are almost exclusively seed eaters, they may have a difficult time on emerging in the spring. Apparently there is no food cache stored away for this period, so the hapless rodents must search for what can be found until the grasses head out again. The method of harvesting grass seed is unique, and once seen will not be easily mistaken. Living as they do in a jungle of tall grass, they are not able to reach the heads nor to climb the slender stems. Instead, they cut off the stem as high as they can reach, pull the upper part down to the ground and cut it again. This goes on until the head is brought within reach. Small piles of grass stems, all cut to an average length, indicate that this is the species which has been at work.

Jumping mice seldom will be seen except when in flight. Then their jack-in-the-box tactics make it almost impossible to determine what they are really like. They are timid, inoffensive little creatures which, if caught, will seldom offer to bite.

Bushy-tailed woodrat
Neotoma cinerea (Greek: neos, new and temnien, to cut ... Latin: cinereus, ashy)

RANGE: Mountainous portions of western North America from Alaska south to central California, northern Arizona and New Mexico.

HABITAT: Found usually in association with the pines of the Transition and Canadian Life Zones; crevices in cliffs and among rock slides are favorite nesting sites.

DESCRIPTION: This woodrat will be recognized at once by its bushy, squirrel-like tail. The several other species in the same range have the usual scanty growth on the tail, so thin as to be almost unnoticeable. This species is large for a woodrat; total length ranges from 12½ to 18 inches. Tail 5½ to 8 inches. The soft, thick fur shows wide variation in color, as might be expected from the great range occupied by this species with its many subspecies. In general it varies from ashy to cinnamon above, to pure white on the underparts. Although the head has the same general shape as that of other woodrats, its appearance is altered somewhat by long, silky whiskers up to 4 inches in length, and extremely large ears. The dark, beady eyes, however, are typical of the genus. The young, from two to six, are born in early summer. This average of four or possibly less, when the breeding habits of all the subspecies are taken into consideration, seems low when compared with other small rodents. The low death rate indicated probably is due not only to this species' secretive habits but to a high order of native intelligence as well.

bushy-tailed woodrat

Many are the names applied to this interesting little animal. "Mountain rat," "pack rat," "trade rat," and woodrat are some of the most common. Several stem from the supposition that when the animal takes an article that suits its fancy, it always replaces it with something which it supposes to be of equal value. Observation of the creature's habits will indicate that these "trades" are entirely by chance. These animals are continually carrying small objects about and often drop one in favor of another more to their liking. The fact is that the most attractive items usually are carried to the vicinity of the nest, and so the scientific name of one of the subspecies is perhaps one of the most appropriate for this industrious collector. This subspecific title is *orolestes*, which translated from the Greek means *oros*, mountain, and *lestes*, robber.

The penchant for carrying away another's property leads to many incidents both comic and tragic. The rats are not at all averse to sharing a prospector's cabin, and during hours when the rightful owner is away at work raise havoc with his possessions. During long winter nights they are no less industrious, and the mysterious sounds of their activities will keep even a sound sleeper awake for hours. Eventually this becomes so exasperating that drastic action is called for. One old prospector told me of a woodrat that had been bothering him for a long time. Traps proved of no avail and finally one night he placed his forty-five on a box beside his bed, together with a candle and matches. During the night he was again awakened and quietly sat up and lighted the candle. There on one of his cupboard shelves was the dim form of the rat. Taking careful aim in the flickering candlelight, he pulled the trigger and hit the animal "dead center." The heavy slug literally blew it apart. Unfortunately it happened to be sitting directly in front of a 5-pound can of coffee. One may assume that without either woodrat or coffee he slept soundly thereafter.

My own experiences with this species have been no less exasperating. When but a youth, my brother and I were quartered in an old bunkhouse one winter. We chose the smaller of the two rooms as being easier to keep warm, and after a thorough clean-up moved in. No rank novices, we wired our watches to a nail driven into the wall and hung our other valuables from a wire stretched across the room. In the morning our socks were missing! Thereafter matters were uneventful for a week. The woodrat would come up through a hole in the corner of the room as soon as the lights were out. All night long it would make trips through the connecting door into the adjoining room and carry away loads of cotton from an old mattress on the unused bed.

Came the week-end and the Saturday barn dance about 3 miles up the canyon. Fresh shirts and trousers donned, coats and vests were taken from the chair backs upon which they had been carefully hung. Behold! One vest front was completely chewed out and carried away, presumably for nesting material. This was the last straw; the creature must be done away with.

On the following night plans were laid with care. Two 5-gallon oil cans were placed in the doorway. This left a narrow passageway just wide enough to accommodate a small jump trap. A piece of newspaper was placed over the trap and the end of the chain wired to the head of the steel bedstead. A short time after the lights were put out, a scratching noise indicated that the animal had come in through the hole. All was quiet until its nose came into contact with one of the empty cans. Then snap! A series of squeaks and the rattle of the chain gave warning that the creature was climbing into bed. As it came in over the head, the wildly excited occupants left by the foot. When the light was struck the rat was sitting in the middle of the bed. A heavy boot soon dispatched it and a semblance of order again returned to the bunkhouse. Strange to say, no more woodrats came in for the remainder of the season.

Although such experiences are the rule when this rat has moved into a dwelling, it is a delightful creature in its native haunts. It is a rim rock dweller; that is, it likes best to build its nest far back in some deep crevice of a cliff. If such a location is not available it may find a protected site in a talus slope or even among the roots of a tree. Usually these natural fortresses are further reinforced by the addition of a pile of sticks and miscellaneous materials piled helter skelter over the nest. The nest itself is quite large, usually a foot or more in diameter, built of the softest and warmest materials at hand. Somewhere adjacent to the nest will be found one or more caches of food against the time when the snows are deep and famine stalks the land. As has been mentioned, the woodrat is usually associated with the pines of the Transition Life Zone and above, and pine nuts are one of its most popular items of food. Acorns, seeds, berries, stone fruits, and some vegetation round out its vegetable diet. It will also eat meat whenever available although, except for insects, shows little inclination to kill its own. With such a varied menu, it seems entirely proper to call this rodent omnivorous.

One of the most characteristic marks of the woodrat's home is a strong, musky odor. This is not an indication of uncleanliness. The animal is most fastidious in its toilette but has this body odor in large measure. A study skin will retain a strong trace of it for many years. Whether it functions for an identification to others of the species is not known, but it could well serve this purpose.

Although classed as a nocturnal animal, the bushy-tailed woodrat is often active throughout the daylight hours. They are not gregarious creatures; yet, since suitable nesting sites may not be found in some areas, other more favored localities often will harbor considerable numbers of the animals. Overhanging ledges may shelter the piles of litter denoting a nest every few feet. In such cases, a well-worn trail will lead from one to the other. This is not an indication that a colony lives there in peace and harmony. These rats are truculent creatures among themselves, and if a stranger should venture into a nest mound, he is evicted with many indignant squeaks and a fearsome snapping of teeth. The interloper seems to know he is out of order and usually leaves the nest at once without more than a token show of resistance. In neutral territory such as a cabin, however, several woodrats may share the area quite peacefully, but to the great annoyance of the human occupant.

The variety of sounds produced by such a group is quite amazing. Added to the usual high-pitched squeaks and patter of running feet are the mysterious rustlings of paper and other objects being dragged about. A peculiar thumping sound indicates a gait which I have never seen but often heard at night. It must be somewhat like the leaping flight of a kangaroo rat, at least it indicates a swift succession of leaps across a flat surface such as a floor or roof. Perhaps the broad surface presented by the flat of the bushy tail is of assistance in this maneuver. Then, like most rodents, the woodrat will thump with its hind legs as an alarm signal. This is perhaps the most noticeable sound of all, for it marks the instant cessation of all activity for every member of its kind within hearing distance. The "ear-splitting silence" that follows this signal literally presses in on one in the darkness.

Muskrat
Ondatra zibethicus (French Canadian word from the Iroquois and Huron Indian word for muskrat. Latin: the odorous substance of the civet alluding to the musk secreted by the muskrat)

RANGE: Virtually all of North America north of the Mexican border. Muskrats are found from near sea level to as high as 10,000 feet above it.

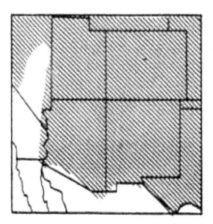

HABITAT: This large rodent can exist only near a permanent water supply which is deep enough to shelter it from its enemies. This may be a lake, a marsh, or a running stream.

DESCRIPTION: A large aquatic rodent whose long, flat tail undulates from side to side when it swims. Total length 18 to 25 inches. Tail 8 to 11 inches. Weight 2 to 4 pounds. The thick, dark brown fur of the upper body is overlaid with brown to black guard hairs. The legs are short but powerful. The front feet are small, but the hind feet are relatively large and partially webbed, with stiff hairs on the edges of the webs and along the sides of the toes. The long, black tail is flattened vertically. It is so scantily haired that it may be said to be naked, but is covered with small scales up to 2 millimeters in diameter. The head is quite similar to that of a vole. The ears are so short as to barely protrude through the fur, and the eyes are small. Average number of young thought to be six per litter. Several litters may be born each year.

The presence of muskrats in a shallow lake or marsh is not difficult to detect. This is their chosen habitat, and here in water about 1½ feet deep, they build their characteristic mounds of rushes and cat-tails. Here they may also be seen on quiet days swimming about and carrying on their normal activities. In much of the Southwest, however, such favorite habitats are few and far between and the muskrats must take their choice, if there be one, between the few permanent streams and irrigation canals. In these altered circumstances they react quite differently; they may often be present in considerable numbers without anyone being the wiser. The change in habits required by this different environment illustrates the great adaptability exhibited by many of our most common mammal species.

The most important requirement of a muskrat is a permanent body of water of a depth sufficient for it to dive into and escape from its enemies. Given this, it will at once set about constructing a home. In a lake or marsh, there is little or no current. In sheltered bays, where wave action is slight, the bottom often will be muddy. In the shallow water along the shore, water plants such as tules and cattails will become established. This is indeed muskrat heaven, for these and other aquatic plants are both their food and building materials. The most edible portions of the plants are the roots and the stem portions which are below the surface of the mud. When one of these choice tidbits has been cut free by the muskrat's sharp teeth, it is carried to some favorite place to be eaten. This may be a mud bar well sheltered by overhanging vegetation from prying eyes, the end of a log projecting above the surface of the water, or perhaps the roof of the "house." The discarded portion of the stem is buoyant and usually lodges among the remaining plants until needed for building purposes.

muskrat

When the muskrat house is being built, a great quantity of this flotsam is piled up until the resultant mound may project as much as 3 feet above the water and be 5 or 6 feet in diameter. The nest is built above the waterline in this half-submerged "haystack." Entrance to the living quarters is by a tunnel which usually starts through the mud a short distance from the base of the house, goes under the edge of the structure, then inclines upward to the nest. Only one entrance is necessary for even if some enemy should tear through the tangle of rushes deep enough to reach the nest, it would take so long that every inmate could easily escape by this submarine route. The house serves one more important purpose in the far north. When the ice lies thick over the marsh and seals this water world away from air, the muskrats can still take short forays under the ice for food and return again to free air, without which no mammal can exist.

Had the muskrat learned to build dams such as beavers construct, the species might very well be near extinction in the Southwest, since such structures would seriously interfere with irrigation. However, since they have accepted conditions as they are, the muskrats do very well for themselves in the shallow streams and irrigation ditches. In fact, their population under the adverse conditions of today is probably far above that of the days before the white man arrived on the scene. Do not assume from this statement that a whole new way of life has been opened up for the muskrat. There has always been a "bank" muskrat that lived in burrows in the stream banks. This fast-water addict has now taken full advantage of the artificial streams that are the forerunners of agriculture almost everywhere in the Southwest. The burrows built into the canal banks seem to be identical with those constructed under natural conditions.

The "bank" muskrat builds three types of shelters, each with a definite and necessary function. These might be called the feeding burrow, the shelter burrow, and the breeding burrow, respectively. The first two are simple in design and have few variations, but the breeding burrow may be extremely complex. If a choice is available, all burrows will be in a bank along the swiftest flow of water, as on the outside of a curve in the canal, for instance. This prevents the entrances from silting shut as they would in the more quiet reaches.

There are two types of feeding burrows. The first and more common consists of a cut made just above water level in the side of a vertical bank. If possible, it is behind a portiere of hanging grass or weeds, so as to be completely screened from view. This is merely a safe place to which the muskrat can take its food and eat without being bothered by enemies. The second type of feeding burrow is more elaborate, consisting of several such chambers along the bank connected by short tunnels. These seem to be community shelters since they are used by several individuals at the same time. The added safety provided by the connecting tunnels seems to be the advantage in this type of dining room.

The shelter burrow not only affords escape from enemies, but may be a sleeping burrow as well. It consists of two tunnels which start at different levels under water and join just before they reach the main chamber, which of course, is above water level. The two tunnels assure an escape route if one or the other is invaded by an enemy. Each muskrat may have several of these shelter burrows. The one used as a sleeping burrow will be furnished with a soft nest of shredded leaves. Cattail leaves are a favorite material for this purpose. Wet, green cattail leaves in a damp underground cavern make a poor bed by most standards, but no doubt, it seems a dry, cozy retreat to the muskrat as it emerges dripping from its underwater tunnel.

The breeding burrows are large and elaborate in design. There is reason to believe they are not always the work of one individual.

They may even represent combined efforts of several generations of muskrats. Often they are a labyrinth of tunnels connecting many nesting chambers, each with a nest of different age. This can be determined by the yellowing of the shredded leaves. As might be expected, there are usually a number of tunnels leading from this maze into the water. A half dozen of these underwater entrance tunnels is not unusual. All this room gives the young a place to exercise before they are able to take to the water.

Young muskrats are surprisingly precocious. They are able to leave the nest when very small, and at 4 weeks of age are weaned and capable of taking care of themselves, although only about one-fourth grown. At this stage, they are peculiar looking little individuals. The fur is still in the woolly stage, dark and bluish in color. The guard hairs have not yet appeared, and altogether they have an unkempt appearance. This rapidly disappears, however, when they leave the burrow. Their progress is so rapid that young born early in the spring are believed to breed during the following fall.

Though ordinarily confined to the immediate vicinity of water, muskrats sometimes are found in amazing places. The urge to travel sometimes influences them to go across country for many miles to some other body of water. They may also become overcrowded in an area so that food becomes scarce and some may leave on that account. It is not uncommon in the Middle West for them to burrow into a farmer's root cellar in early fall and spend months in this haven of warmth and good food before they are discovered. Floods may carry them many miles away from established haunts and leave them stranded on high ground when the waters recede. A muskrat found in this predicament is not an animal with which to trifle. If it cannot escape by water, it will probably elect to make a stand. The long, sharp incisors are formidable weapons indeed, and any enemy, including man, had best allow judgment to become the better part of valor.

The tracks of muskrats are so characteristic that they cannot be mistaken for those of any other animal. Strangely enough they resemble to a striking degree those of certain types of extinct reptiles called dinosaurs. The tracks of the two small front feet are close together and overlapped somewhat by those of the larger hind feet. Between the tracks is the sinuous trail left by the sharp-edged tail.

Beaver
Castor canadensis (Latin: a beaver ... from Canada)

RANGE: The beaver, like the muskrat, can be found almost everywhere in North America north of the Mexican border.

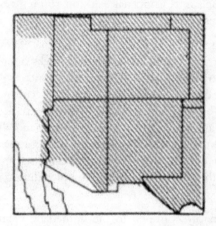

HABITAT: Near any water supply of enough volume, with or without damming, to provide security for a beaver family.

DESCRIPTION: The largest North American rodent; further distinguished by having a broad flat tail. Total length 34 to 40 inches. Tail 9 to 10 inches. Weight from 30 to 60 pounds. In color the beaver varies from a deep, rich brown in the northern states to a much paler shade in desert regions of the Southwest. The soft, rich underfur is partially concealed by coarse, rather stiff guard hairs. The brown color of the upper parts shades to a chestnut under the belly and on the inner sides of the legs. The forefeet are small with well developed claws. They appear naked but have a scanty cover of coarse hairs. The hind feet are large and webbed, and are similarly covered with a few coarse hairs.

The body of the beaver has somewhat the appearance of a kangaroo in that the rear portion is heavy and appear overdeveloped in comparison with the more stream-lined head and forequarters. Much of this impression is gained from the heavy, flat tail which is thick and muscular at the point where it joins the body. One of the most useful appendages possessed by any creature, the tail is paddled-shaped horizontally and about an inch thick in the middle, tapering to thin edges and tip. It appears naked, but is covered with scales.

The young, averaging four in number, are born in the late spring and, although they are soon able to take care of themselves, the family remains together for most of the year.

Indications of beavers in an area are their dams or the distinctive stumps left by their tree felling. Beaver tracks are seldom found. Although this aquatic animal often leaves the water, and may go a considerable distance overland, its tracks usually are obliterated by the passage of the heavy rump and the dragging tail.

The beaver, perhaps as much as any other factor, was instrumental in opening up western America to civilization. Even before the Thirteen Original Colonies had become firmly established along the eastern seaboard, venturesome men were working westward in search of more beaver to supply the ever-increasing demand for this soft-rich fur. Industrial empires were founded on this traffic in skins which came from as far west as the Mississippi River. By the early 1800's, the trappers had penetrated to the Rocky Mountains, and in 1806, upon the return of the Lewis and Clark Expedition from the Pacific Northwest, they swarmed to the headwaters of the Missouri River system. Prior to this, the Southwest had been given little attention by the fur industry. It was considered an inhospitable region, inhabited by hostile Indians, and with a few settlements of Spanish colonists who, up to that time, had actively resisted the intrusions of the more aggressive Americans. However, by the year 1820, relations had improved to such a degree that a few of these hardy individuals were trapping on the headwaters of the desert rivers. Later, their activities spread to include the entire length of these remarkable watercourses.

These were the Mountain Men, a hard-as-nails breed of frontiersmen in a class by themselves. In the period from 1820-1854, when a large part of the Southwest became part of the United States through the Gadsden Purchase, they roamed the plains and mountains of the American Desert. Their roster includes such legendary figures as Bill Williams, Pauline Weaver, Kit Carson, and James Pattie. Their argosy was a quest for the rich, brown beaver pelts which were a golden fleece indeed when presented to the fur traders in far-off St. Louis. In time, their moccasined feet beat a broad path across the western plains—a path then known as the Santa Fe Trail, but identified today as U.S. 66, the "Main Street of America."

Today, many of the streams which supported beaver colonies in the desert places have vanished entirely, and others have been so effectively harnessed for irrigation and power that there is no place for beavers in them. In the higher mountains, however, there are many streams remote from civilization where clear ponds still sparkle in the sunlight, and the splash and dripping of busy beavers can be heard on quiet, summer evenings. Because beavers quickly become established under any conditions which are at all favorable, they have been reintroduced into numerous places where they had been extinct for many years. Usually this is good conservation practice, but under some conditions, it may prove a mistake. Ecologically speaking, beavers probably are the most important creatures in any animal community of which they are members. This is because these busy engineers not only impose a tremendous drain on the surrounding area for material, very often they also radically alter the character of the terrain to fit their own needs.

beaver

The life history of the beaver is one of the most interesting of all mammals. It has been studied for centuries by naturalists in both the New and Old Worlds, for the beaver, with but few differences, is native to both. All this study and observation notwithstanding, the habits are still only partially known. This is because the beaver is mainly a nocturnal creature which spends most of its daylight hours in the concealment of a lodge or burrow. Then, too, in the northern latitudes where the ponds are covered with ice throughout the long winters there is little opportunity to observe this phase of its existence. There is but one species of beaver in North America but about two dozen subspecies. The northern types and those which live in the mountains of the Southwest seem to be dam builders who live in beaver "lodges." Those which inhabited rivers of the lower desert were mostly "bank" beavers which lived in burrows in the banks of streams. This latter type is rare today.

Perhaps the best way to understand the ecological importance of the beaver is through watching the rise and decline of a typical colony. Picture if you will a small, shallow stream flowing gently down a narrow valley in the mountains. Bordering the low banks is a thicket of alders. Back of them a thick growth of aspens extends to the edge of the valley and mingles with the spruce trees on the slope. Down this slope comes a young male beaver at a clumsy gallop, his broad tail striking the ground with an audible thump at every lope. This emigrant has struck out for himself because the colony to which he belongs has become crowded. He finds the stream and, since the water is too shallow to conceal him, crouches under an overhanging bank until darkness falls.

As soon as it becomes completely dark, he hunts for a suitable place to build a dam and soon finds a site to his liking. On one side of the stream a thick clump of alders projects from the bank, and on the other a water-soaked log is half buried in the bottom of the creek. From these anchor points, he begins his dam, building toward the middle from each side. The work calls for a great deal of the alder brush to be cut and sunk in the bed of the stream. There it is weighted down with rocks and mud until secure. Additional brush is brought and interwoven with the first; gradually the structure grows until in a few days it converts the stream into a quiet pool deep enough to hide the beaver, should an enemy appear. As the water rises it covers the bases of the alders, which begin to die in the pond.

The beaver next turns his attention to building a lodge. Selecting a point to one side of the current entering the pond, he begins as he did with the dam by sinking brush to the bottom and weighting it down with rocks. As he builds, he cleverly fashions several underwater entrances to the house that will be. When he has finished, the house projects several feet above the water, and the materials are so thoroughly interlaced and plastered that even the most determined enemy would despair of gaining entry to the living room. Debris from the construction has floated downstream to become lodged in and on the dam, making it more secure and watertight that it was when first built.

With the dam and the lodge both completed, the next task is to collect a food supply for the following winter. This is carried on intermittently during the autumn. It consists of cutting down aspens, whose bark the beaver dearly loves, sectioning the branches and small trunks into pieces which may be handled conveniently, and dragging them to the pond. Once in the water, they are weighted down and will remain in good condition for a long time. The beaver is joined in this task by a female which has also migrated from an overcrowded colony. Two need more food than one, consequently their trails begin to head a little farther into the aspen forest as they work through the crisp autumn nights. These trails converge as they leave the forest and approach the pond, and end in a few well-developed mud slides that enter the water. Constant traffic of the wet beavers leaving the water keeps the slides moist and slippery.

As winter settles in on the mountains, a thin skim of ice begins to form on the edges of the quiet water on cold nights. Then one night it freezes completely over. This causes the beavers no inconvenience at all because if on one of their underwater excursions they should wish to surface for air, they have but to swim to a shallow place with firm bottom, and with one quick lift of their powerful muscles break a hole through the ice with their backs. They can break surprisingly thick ice in this way. The beavers live in comfort and plenty throughout the winter. The living room of the lodge has been furnished with comfortable beds of the cattails that have already become established along the edge of the pond. The lodge, although tightly built, still admits enough air for the beavers and food is stored in plenty on the bottom of the pond. As the bark is gnawed from the aspen branches, the bare poles are added to the bulk of the house or used in further construction of the dam. Before long, the mild southwestern winter merges into spring.

In late spring the beaver family is considerably increased by the arrival of four miniature beavers. They weigh but 1 pound each at birth and are fully furred. At this time, the father is ostracized and the mother and her young live together in the lodge. When the young are about 3 weeks old, they take to the water for the first time. They quickly learn the beaver method of swimming; this is to kick with the hind feet and let the forelegs trail loosely alongside the breast, using the flat tail both as elevator and rudder. The young beavers are called kits, and indeed are as playful as true kittens can be. It is most amazing to watch them cavorting about in the water with as much ease as youngsters of other mammals do on dry land. As autumn nears, this play is exchanged for the sterner duties of existence, and the young take their places as adults of the family.

Fifty years pass. As the colony increases the dam must be made larger, new lodges must be built; and when the trails to the aspen forest become too long, canals are dug part way out to lessen the hazards which may befall the beaver on dry land. The pond gradually silts up to higher and higher levels until at last it is full of black, fertile soil. All of the aspens within reach are finally cut down and the hungry beavers turn to the resinous bark of the spruces. Finally the struggle is given up. The beavers migrate to a new location, and the following spring a freshet tears out the center of the dam. Now the pond is gone. With it are gone the trout that played in its depths, and the teal that rested there on their way south. In its place is a beaver meadow, a grassy park in the center of the spruce forest with spring flowers spangling its green surface. Aspens are already beginning to crowd in about its edges, and the creek is cutting deeper into its soft soil with every spring. Before long heavy erosion will begin to take its toll, and some day in the future a male beaver will again come galloping awkwardly down the slope.

The changing conditions which such a cycle bring about are almost impossible to evaluate. At least three climax types of environment are represented: those of the alder thicket, the beaver pond, and the beaver meadow. In a graphic fashion this cycle illustrates what is going on in Nature continually, more slowly perhaps, but just as surely.

Porcupine
Erethizon dorsatum (Greek: to irritate in allusion to the quills and Latin: pertaining to the back)

RANGE: Most of North America north of the Mexican border. Notable by their exception are the south central and southeastern United States.

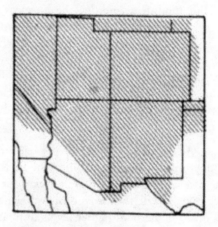

HABITAT: Usually associated with conifer forest, yet may sometimes be found miles from any forest. An inhabitant of all life zones up to timberline (Arctic-Alpine).

DESCRIPTION: A black to grizzled black and yellow creature covered with quills. Total length 18 to 22 inches. Tail 7 to 9 inches. Weight 10 to 28 pounds. Body short and wide; supported by short bowed legs. Tail heavy and muscular, armed with short slender quills. Head small with dull eyes and long black whiskers, but with short ears. The incisors are extremely large and are of a bright, rich yellow color. The quills are shortest on the face and reach their greatest length near the middle of the back. Often they are nearly hidden in the coarse, seal brown to black underfur. The long guard hairs are also seal brown close to the body, but change to a rather sere yellow at the tips. Only one young is brought forth each year in a den among the rocks, or sometimes in a hollow log. The young are among the most precocious of any mammal.

The porcupine in North America is considered as belonging to but the one species *dorsatum*, although there are seven subspecies. The most common subspecies found in the Southwest is *epixanthum* (Greek *epi*, upon, and *xanthus*, yellow), sometimes called "yellow-haired" porcupine. The porcupine is unique among North American mammals in bearing the sharp quills which are perhaps its most interesting feature. Certainly they are responsible in large part for the unusual life history of this misunderstood animal.

Quills are no more than greatly modified hairs, and in sorting through the various types of pelage on a porcupine's back, a few examples will be found which are intermediate between the hair and the quills. This does not mean that coarser hairs gradually turn to quills. Each follicle produces hair or quill, as the case may be, for the life of the animal. A quill consists of three well-defined parts: a solid sharp tip usually black in color; a hollow shaft, which is white; and a root similar to that of a hair.

porcupine

The sharp tip is smooth for a fraction of an inch, but from this point on, it is covered with a great number of closely appressed barbs. These can be felt by rubbing the quill the "wrong" way between thumb and forefinger. It has been found that these barbs flare away from the surface, when the quill is immersed in warm water. It seems natural that they would do the same when embedded in warm, moist flesh. At any rate, quills are always difficult to extract, and if left in the victim they penetrate ever more deeply until they may pierce some vital organ and cause death. In other cases, they have been known to work entirely through body or limb and emerge on the opposite side. This is due to muscular action of the victim, some movements tending to force the point farther, the barbs at the same time effectively preventing any retreat.

Below the barbs the tip of the quill flares to join the shaft. Pure white and opaque, this portion is used by Indians to form decorative bands of quill work on the fronts of buckskin vests and jackets. This part is also hollow, and before removal of a quill from the flesh is attempted, a little of the end should be cut off. This collapses the shaft and makes extraction somewhat more easy, but very little less painful. Actually there is little excuse for a human to become involved with one of these mild-tempered creatures, but sometimes dogs are badly hurt in encounters with them.

The root is the portion by which the quill is attached to the body. Although it is a common belief that the porcupine can "throw" its quills, the truth is that the root portion is extremely weak and the quills are easily withdrawn from the body when the barbed tip is driven into an enemy. In fact, any violent movement of the animal may dislodge quills, even though nothing has touched them. There are several well-authenticated accounts of quills having been flipped for several feet in this way, but in each case, it was entirely accidental and through no conscious effort of the porcupine. In other words, the armament of this slow, awkward creature should be considered strictly defensive in every respect.

Like the skunk, which can also defend itself most effectively, the porcupine has little apparent fear of its enemies. When threatened with violence it simply brings its head down between the forelegs and turns its rump toward the attacker. With hair and quills erect it resembles a soft furry ball. Appearances are seldom more deceiving! The guard hairs half conceal a spot on the back where a whorl of long quills radiates out in a large "cowlick." Should any enemy touch these long guard hairs, the muscular tail is thrashed vigorously about in an effort to drive the somewhat shorter but equally keen-pointed tail quills into the attacker. With every attempt at attack from another angle, the porcupine turns so as to present its rump to the enemy. There is one Achilles heel, however, in this otherwise almost perfect defense. It is the unprotected underparts, which at times of danger are always kept pressed against the ground or against a tree trunk. A few carnivores, among them the mountain lion and the fisher, are known to kill the porcupine by flipping it over on its back and tearing it open. Even these large predators seldom escape unscathed, however, and both lions and fishers are known to have died from the effects of quills accidentally taken into the digestive tract.

To those who have heard that porcupines live only on bark and always girdle the host trees, it may come as a surprise to find that this is only partly true. Although "bark" is eaten to some extent throughout the year, it is seldom the main diet. When a great deal is taken from one tree, it is gnawed off in an aimless pattern which may or may not girdle the tree. During the spring and summer, a porcupine becomes a browser on tender leaves and twigs in the undergrowth. In autumn and winter, it feeds more on mistletoe and pine needles than on bark. With its low reproduction rate, there is little danger of it eating up our forests, unless its natural enemies are removed.

Northern pocket gopher
Thomomys talpoides (Greek: thomos, a heap and mys, mouse. Latin: talpa, a mole)

RANGE: From northwestern United States and southwestern Canada to as far south as northern Arizona and northwestern New Mexico.

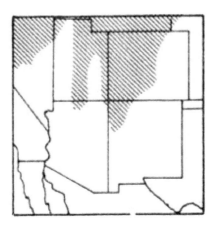

HABITAT: Soft loam in the open places in the high mountains. Seldom found below 8,000 feet, but up to elevations of over 13,000 feet in New Mexico.

DESCRIPTION: The characteristic mounds of earth built up by this group of burrowing rodents are usually the best indication of their presence. The northern pocket gopher is of medium size. Total length 6½ to 9½ inches. Tail 1¾ to 3 inches. It is usually gray in color with darker patches behind the rounded ears. Eyes and ears are small. The short tail has a bare, blunt tip. Front claws are long and curved. The entire body is well muscled and gives an impression of power. Average number of young thought to be about four. At the high elevations at which this species lives, the young are not seen until rather late in summer.

northern pocket gopher

The northern pocket gopher is one of the hardiest rodents on the North American Continent. Even so, it would not be able to survive the climate of the inhospitable regions it sometimes inhabits were it not for the fact that is spends almost all its life underground. This creature does not hibernate, but continues busily at the task of searching out food when most other subterranean dwellers are curled up fast asleep in their cozy nests. Why the gopher should continue working, while its ground squirrel cousins sleep, is hard to say. It would seem that it has the same opportunities to lay on fat for a winter's rest. The chief reason seems to be that the bulbs and roots upon which it feeds are always available so long as the gopher keeps extending its underground workings. On the other hand, the ground squirrels, which gather their food aboveground, are cut off from this supply as soon as cold weather drives them to shelter.

The pocket gophers are much alike. There are three genera and a considerable number of species represented in the Southwest but, except for variations due to climate and terrain, their habits are similar. Burrows usually are constructed in deep loam or alluvial soils. These tunnels seem to follow an aimless pattern. Their course is marked by mounds of earth thrown out of the workings at irregular intervals. When the gopher is engaged in throwing out this excavated earth, the entrance to the tunnel is left open until the job is completed, then tightly plugged to prevent enemies from entering. The tunnels themselves are rather small in diameter, considering the size of the gopher, for if it wishes to retrace its steps and there is no gallery near at hand in which to turn around, it can run backward almost as easily as forward. There are usually numerous rooms excavated along the course of the tunnels. In one is a warm nest constructed of grass and fibers. Others are utilized for storage rooms and at least one is reserved as a toilet, thereby keeping the rest of the workings sanitary. When the ground is covered with snow the northern pocket gopher especially is quite likely to extend its activities aboveground. Here it builds its tunnels through the snow and often packs them tightly with earth brought up from below. This remains as earth casts, when the snow melts and forms a characteristic mark of its presence.

Chief foods of pocket gophers are the bulbs, tubers, and fibrous roots encountered in the course of their diggings. Whenever an especially abundant supply is found, the surplus is stored away as insurance against the time when future excavation produces nothing. Gophers also eat leaves and stems whenever available. Some plants are pulled down through the roof of the tunnel by the roots, and some are gathered near its mouth, although these trips "outside" are fraught with danger. Coyotes, foxes, and bobcats all are willing to chance an encounter with this doughty little scrapper for the sake of the tasty meal he will furnish.

Little is known of gopher family life. For the most part, they are solitary individuals, avoiding others of their kind. At breeding time, however, they may travel some distance across country to find a mate. These trips usually are carried out under cover of darkness. The young average four in number. They are born late in the spring and do not leave to make their own homes until early autumn.

Physically the gopher exhibits a striking adaptation to its way of life. The fur is thick and warm. It keeps soil particles from working into the skin at the same time it protects the wearer from the chill of his underground workings. The heavy, curved front claws are admirable digging tools. In especially hard soil, the large strong incisors are also pressed into service for this purpose. To remove the dirt from the tunnel, the gopher becomes an animal bulldozer. The front legs are employed as a blade pushing the soil, while the powerful hind legs push the body and load towards the nearest tunnel opening. The pockets from which this creature gets its common name are never used for hauling earth. They are hair-lined pouches located in each cheek and utilized for carrying food to the storerooms. There they are emptied by placing the forefeet behind them and pushing forward. Last, by virtue of its location, but certainly not least in usefulness, is the short, almost hairless tail. It is used as a tactile organ to feel out the way when the gopher runs backwards through the tunnels. In some respects, it is of more use than the eyes although the gopher uses these too, as can be attested by the quickness with which it detects any movement near the mouth of its tunnel.

The gopher's place in Nature seems to be akin to that of the earthworm. By turning over the soil, the gopher enables it to more readily absorb water and air. At the same time, fertility is increased by the addition of buried plants and animal matter. This is indeed a fair exchange for the plants it destroys in its quest for food.

CARNIVORES
Including the Insectivores and Chiropterans

This group is distinguished from other animals by having canine teeth in both jaws. The function of these teeth is to catch and hold other animals, for carnivores are the predators. This is the most highly developed branch of the animal world and reaches a peak of specialization in man who, while lacking some of the physical qualifications of the other predators, has developed a brain which has enabled him to gain and keep ascendancy over all other animals. Considered with the group in this book are two other orders, the Insectivora and the Chiroptera. These orders embrace the mammals in North America that live principally on worms or insects rather than on other mammals. They are the shrews and bats, respectively.

Since carnivores are the hunters rather than the hunted, they enjoy far greater mobility than, for instance, the rodents. It is not necessary that they have a burrow in which to escape the attacks of other animals, for it is unusual for them to prey upon each other. Most of the predators remain in one area only from choice or, in the case of adult females, in order to rear the young. Few of them hibernate; bears and skunks do spend a considerable time during the cold weather in a torpor, but it is an uneasy sleep at best, as anyone who has disturbed these animals at this time can attest. As far as the Chiroptera are concerned, some species of bats hibernate and others migrate to a warmer climate to spend the winter. Since most of the predators are active all winter, while many of the rodents are in hibernation, this can be a period of famine for carnivores. At the same time, it is a season of increased danger for those species which are still active and upon which these predators prey.

Because these hunters are continually stalking other animals, their habitats are as varied as those of their quarries. Thus, the mountain lion is a creature of the rimrock, where he can most conveniently find deer browsing on mountain-mahogany; while his smaller cousin, the bobcat, stalks smaller animals in the slope chaparral. The wild dogs hunt plains and brushy country for ground squirrels and rabbits. In the weasel family we find the marten in the treetops pursuing squirrels, the weasel hunting mice in the meadow, and mink and otter pursuing prey near to or in the water, Some species, such as the bears, are omnivorous and may be encountered almost anywhere that a plentiful supply of food of any kind can be found. Practically all of the species, excepting bats and skunks, can be considered diurnal as well as nocturnal, but the majority are most active during the hours between dusk and sunrise.

Since the carnivores' purpose in Nature's scheme is to control the vegetable eaters, it follows that each predator must be somewhat superior, either physically or mentally, or both, to the species upon which it preys. The associations between pursuer and pursued may be casual with species such as the coyote, which preys on a great number of smaller species, or they may be sharply defined as with the lynx, which in certain localities depends almost entirely upon the snowshoe hare for food. The apparent ferocity with which some predators will kill, not only enough for a meal, but much more than they need, cannot as yet be explained. This habit is most pronounced in the weasel family. It may be that more than ordinary control is called for in the case of their host species, rodents in most cases. Whatever the reason, this wanton killing has not upset the balance which these species maintain. Man, the most ruthless and intelligent predator of all, is the only species which has been successful in exterminating others.

The predators hold a favored place in the esteem of most naturalists. At first, sympathy for the weak and indignation against the strong are perfectly natural human feelings. As the necessity for control and the wonderful way in which Nature attains a balance becomes apparent, the role of the predator becomes more and more appreciated by the student.

Mountain lion
Felis concolor (Latin: a cat of the same color; referring no doubt to the smooth blending of the body coloration)

RANGE: At present, mostly confined to the western United States and Canada, and all of Mexico south to the southern tip of South America. There are a number of mountain lions in Florida, and persistent reports indicate that they may be making a comeback in a number of other Eastern States.

HABITAT: As the range indicates, habitats vary widely. Mountain lions in the Southwest show a preference for rimrock country in the Transition Life Zone or higher, but they are often seen in all the life zones.

DESCRIPTION: A huge, tawny cat with long, heavy tail. The long tail is a field mark identifying the young, which, having a spotted coat, otherwise resemble young bobcats to some degree. Total length 72 to 90 inches. Tail 30 to 36 inches. Weight 80 to 200 pounds. Color may vary from tawny gray to brownish red over most of the body, the underparts being lighter. The head and ears appear small in proportion to the lean muscular body. The teeth are large, the canines being especially massive. Like most members of the cat family, the mountain lion has large feet with long, sharp claws. The tracks show the imprint of four toes together with a large pad in the center of the foot. The young may be born at any time of the year. Only one litter is born every 2 to 3 years, and the average number of young is three.

Probably no species of mammal in the New World equals the mountain lion in farflung distribution. From the Yukon to Patagonia, this elusive carnivore can still be found in considerable numbers in spite of aggressive campaigns against it. In the United States, it is the chief representative of the wild cats, a group noted for fierce and predacious habits. Fortunate indeed is the person who sees one of these great felines in the wild. This may not be as difficult as one might imagine because mountain lions often travel through comparatively well settled areas. It is especially possible in the Southwest, for the four-State area covered by this book contains the heaviest population of mountain lions in the

United States. However, the comparative abundance of this carnivore has not resulted in a better understanding of it. The mountain lion is still one of the least known and most maligned creatures of our times.

mountain lion

The Mexicans know this cosmopolite as "leon." In Brazil it is called "onca." Perhaps the most distinguished name, and rating as the first in New World history, is "puma," given it by the Incas. Early American settlers of the east coast called it "panther," "painter," and "catamount." In the northwestern United States, it is known as "cougar" and in the Southwest, as mountain lion. Although there is but the one species *concolor*, there are a number of subspecies. About 15 are now recognized, most of them geographical races and not markedly different from the species. Four of these subspecies are found in the four States with which we are concerned. One of the most interesting is *hippolestes* which inhabits the State of Colorado. Translated from the Greek this is "horse thief," an appropriate epithet indeed for this ghostly marauder. As might be expected from their vast distribution, the several subspecies have a tremendous vertical range. In the Southwest they are found from near sea level in southwestern Arizona to the tops of the highest peaks in Colorado.

In the more than four centuries that have elapsed since the white man first set foot on soil of the New World, a great mass of folklore concerning the mountain lion has accumulated. Half fact, half fiction, these tales have been repeated from one generation to another and few details have been lost in the telling; indeed, in most cases, several have been added. Most common are those which describe its fierceness and its attacks on man. In the main, these tales are lurid and convincing, but they do not stand up under scientific scrutiny. It is true that such attacks have occurred; one of the most recent and best verified was that on a 13-year old boy in Okanogan County, Washington, in 1924. It resulted in the death and partial devouring of the unfortunate youngster. Yet sensational as this incident was, it resulted in publicity far out of proportion to its importance. In fact, articles concerning this case are still appearing at intervals. The truth of the matter is that very few authentic cases of mountain lion attacks upon humans have ever occurred in the United States, and that most of these *could* have been caused by the mountain lion's being rabid. Certainly such attacks are not typical behavior of the normal animal. As far as man is concerned, the lion will take flight whenever possible, and even when cornered it is not nearly so pugnacious as its little cousin, the bobcat.

Other stories about the mountain lion often emphasize the bloodcurdling screams with which it preludes its stalk of some unfortunate person deep in the forest. The facts are that there is no reason to believe that lions cannot or do not scream, but most authorities agree that such vocal expressions are most likely to be made by an old male courting his lady love or warning away a rival. The cats are creatures of stealth and cunning that creep upon their prey as noiselessly as possible. Lions would hardly announce their presence with the sort of screams with which they are credited. It seems safe to say that at least 90 percent of these alleged screams can be traced to owls or amorous bobcats. Oftentimes these sounds have been linked to large tracks found in the vicinity as proof that a mountain lion was in the area. This has led one author to remark that "the witness usually is unable to distinguish the track of a large dog from that of a mountain lion." In addition, the infrequent screams made by captive mountain lions indicate that such sounds in Nature would be far from spectacular. They consist of a sound that is more like a whistle than the demoniacal wail so often ascribed to the wild animal.

Many stories are told of a person, usually a pioneer ancestor, who has been followed by a mountain lion. In most cases this person has returned to the area suitably armed and with witnesses who found tracks of the beast together with those of their friend. Strange to say, such incidents are not at all uncommon. They have been recorded and verified a number of times. In these cases the animal often has made no effort at concealment but has followed the person quite openly. Despite this boldness it seems there is no sinister motive, merely a naive and surprising curiosity on the part of the big cat as to what kind of creature man is. It is most unfortunate that so little data have been recorded in these instances, yet this is quite understandable under the circumstances.

Finally, in most stories there is only one size of mountain lion— big! As the story makes its rounds the lion never gets smaller; it invariably grows larger. Somehow the records have missed all these really big lions. Any lion which measures more than 8 feet in length and 200 pounds in weight will be an extremely large, old male in the record class. The average will be much smaller. Statistics show most lions to be 5 to 7 feet in length and 80 to 130 pounds in weight for adult females, and 6 to 8 feet in length with weights of 120 to 200 pounds for adult males. Errors in estimating the size of these big cats are easily accounted for. In the first place the lion is a long, low, sleek creature that gives an impression of being longer than it is. Too, its size is unconsciously exaggerated by many people who are impressed with its tremendous power and agility. Many of its feats of strength seem impossible for an animal so small. Lastly, its tanned hide may be available for measurement. Actually this proves nothing; hides often are stretched 2 feet or more at the time the animal is skinned, and tanning does not shrink them appreciably.

None of the above is meant to detract in any way from the reputation of the mountain lion or its place in American folklore. It is the third largest predator in the Southwest, being exceeded only by the jaguar and the bear in size, and surpassing them both in agility. In the past, it has been feared and hated by those whose herds and flocks have suffered from its depredations. Their efforts to exterminate it have resulted in grave biological problems at times, but in the light of more advanced study it seems probable this big carnivore will be spared in the future to keep its rightful place in our wilder areas.

The mountain lion "goes with the deer"; that is to say, its function is to keep deer in check so that they will not eat up their range and starve to death. Though at first glance such a possibility seems out of the question, this has become a serious problem in recent years. It will be further intensified as suitable deer range becomes more restricted with the advance of civilization. Another function of the mountain lion-deer relationship is to weed out the diseased and inferior individuals so that the deer herd will remain healthy and up to good physical standards. It may be argued that the same end is reached by hunting, and so it is, with one major exception. The nimrod, intent on a fine trophy head, takes the buck in the prime of life, a time when he should be sireing the herd of the future. The cougar does not consciously select its victims; it takes the most easily caught, thus leaving the wisest and healthiest survivors as breeding stock.

Though deer are the lion's preferred food, many other species of mammals are preyed upon when deer are scarce. These range in size from the smallest rodents to animals as large as elk. Among the more unlikely species recorded are skunk and bobcat. The lion also has the dubious distinction of being one of the chief predators of the porcupine. Dining on this last species is fraught with danger, however, because no matter how expertly the carcass is removed from its spiney covering, a few quills will penetrate the flesh of the diner. Little prey other than mammals is ever taken. Birds are not easily caught by such a large animal and, although it does not shun water, it is poorly equipped to take any form of aquatic life. The mountain lion will not eat carrion except under the most dire circumstances and prefers food that it has killed itself.

There are two principal methods by which the mountain lion catches its prey. The stalk and pounce technique of the common house cat is most effective in brushy country where the low crouch of the lion places its bulk behind the close ground cover. With tip of tail twitching, it creeps forward until a short run and spring, or the spring alone, will carry it to the front flank of the unsuspecting victim. If the neck of the hunted is not broken by the impact of the heavy body, the sharp claws or massive canine teeth are brought into play to rip the jugular vein and end the struggle. In the other method of hunting, the lion chooses a ledge above a game trail and simply waits there until some animal to its liking passes below. The weight of its body usually is sufficient to bear the victim to the ground and it is soon dispatched. Mountain lion studies in California have determined that in hunting deer the animal will catch one in every three attempts. It has been estimated that in an area of heavy deer population each mountain lion will kill one each week. It is of interest to note that in many places in the Southwest deer are on the increase, indicating the need for more predators to keep down their number.

Since the mountain lion has few enemies, its reproduction rate is low. Two to four kittens are born in each litter, but usually at 2- to 3-year intervals. Dens are sometimes located deep among the rocks; others may be no more than a grass nest in the brush on a rocky ridge. Like domestic kittens the young are born blind. They have an interesting color pattern at birth, a strongly spotted coat and a faintly ringed tail. This completely disappears when they are about half grown, leaving them with the tawny reddish coat which blends so well with their surroundings. They mature at about 2 years of age; beautifully evolved killers which must be admired by everyone who has come to understand the methods by which Nature regulates the animal world.

Bobcat
Lynx rufus (Latin: name of animal, and rufus, reddish)

RANGE: Common throughout much of the United States and Mexico. Found throughout the Southwest.

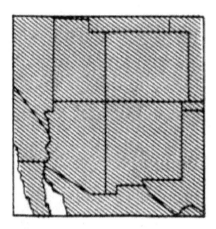

HABITAT: This common species is found in all zones wherever there is sufficient cover to hide it.

DESCRIPTION: A bobcat distinguished from the lynx by having small ear tufts, a more rufous color, and a black band which crosses only the upper surface of the tail tip. Total length 30 to 35 inches. Tail 5 inches. Weight 15 to 30 pounds. This is a chunky animal with long, muscular legs and large feet. The sides of the face are heavily streaked with black, backs of ears dark, coat generally tawny to rufous above, underparts lighter. Dark spots rather prominent throughout coat, insides of front legs often barred with darker color. Young from two to six, usually born in early spring; only one litter per year.

These are the most common wild members of the cat family in the Southwest. Their distribution over the United States takes a strange pattern, inasmuch as they are not found in several of the midwestern and southeastern States, and in a large area in central Mexico. In all there are a dozen subspecies of *Lynx rufus* in North America. They are tough little predators, among the last to retreat before the advance of civilization. In fact, they may often be found on the very fringes of our larger cities, existing on the rats that infest the city dump.

In the wilder areas, which are the bobcat's appropriate home, its tracks are distinguishable from those of the larger *Felidae* only by their smaller size. Like the larger members of the cat family, it is equipped with a set of strong retractile and extremely sharp claws. Although there are five toes on each front foot and only four on the hind feet, the tracks of both feet are similar. This is because the fifth toe, corresponding to our thumb, is so high on the inside of the foreleg that normally it does not touch the ground. During normal travel the claws are always in the retracted position and never show in the tracks. All native cats have a tendency to place the hind feet in the tracks left by the front feet, so that in effect each track is a double print. This may be one of the reasons a cat's approach is so silent!

bobcat

Bobcats have numerous traits in common with their relative, *Lynx canadensis* (not treated in this book because of its extreme rarity in the Southwest), but are more versatile in their dietary tastes. While the lynx is sufficiently dependent on the snowshoe hare that its population corresponds closely in fluctuation with that of its "host," the bobcat has a much less discriminating appetite. It also loves snowshoe hares and rabbits, but takes various other mammals as opportunity offers, and ground-living birds. Bobcats will even eat carrion, but prefer fresh meat. They are reliably reported to eat porcupines, young pronghorns, deer, and sheep, both bighorn and domestic; and they sometimes kill adult deer, although this is a difficult and dangerous proceeding. Usually a kill is at least partially covered with debris, and the cat will return at least once to feed again on it.

Though bobcats are the least spectacular of our native cats they are the most numerous and evenly distributed. Thus collectively they may be of more importance in Nature's master plan than we realize. Their role may even increase in importance as time goes on, because of the increasing scarcity of the larger cat species.

Red fox
Vulpes fulva (Latin: a fox ... fulva, meaning deep yellow or tawny)

RANGE: Found throughout most of North America north of the Mexican border. Exceptions in the United States are areas in the southeastern and central States and desert portions of the Southwest.

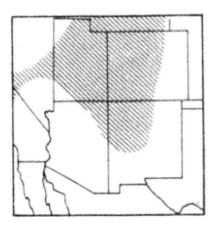

HABITAT: In the Southwest these foxes are restricted to wooded areas of mountains. They usually are found in the Transition Life Zone or higher.

DESCRIPTION: About the size of a small dog, having a bushy tail with white tip. Total length 36 to 40 inches. Tail 14 to 16 inches.

Weight 10 to 15 pounds. Besides the type, this fox has at least two well-defined color phases with many intermediate forms. These will be considered separately. A typical western form of red fox will be more yellow than red. The brightest red will be a rufous median line running down the back. This fades to an ochre yellow along the edges and grades to the lighter yellow of the sides. The tail is usually dark yellow with black guard hairs and always a white tip. The underparts are light yellow to white. Fronts of feet and lower legs and backs of ears are always very dark to black. The underfur is lead-colored. The head is small with large ears, yellowish eyes having elliptical pupils, narrow nose and jaws. The young, four to six in a litter, are born early in the summer and but one litter is produced each year.

The western form of red fox might more aptly be named the "yellow" fox, since it is definitely more yellow than red. To add to the confusion, the gray fox, *Urocyon cinereoargenteus*, of the West usually has more good red in its coat than the red fox. However, the gray fox is a denizen of the desert and will not often be found at elevations preferred by the red fox. In addition, its tail is tipped with black; this definitely separates the two species at a glance. The differences of color phases within the red fox group are more pronounced and have led many people to consider them separate species. The two most distinct types of these varieties are known as the "cross" fox and the "black" or "silver" fox.

The term "cross" fox refers neither to the disposition of the animal nor to its being a hybrid variety, although it often is cross or mean and is not a hybrid. It alludes to the dark cross on its back. This is formed by a dark to black median line crossing at right angles to a dark band that traverses the shoulders. Its effect is increased by considerable amounts of gray and black mixed with the normal yellow color of the sides. The long hairs of the tail are yellowish gray to black, the general effect being dark but, as with the type, the tip is pure white. As might be expected, there are many gradations between this color phase and the type, some of them being among the most striking and beautiful foxes in the world.

The "black" or "silver" fox is a melanistic form of the red fox. In the most striking form it is a smooth shining black, the general sombreness of its coat being relieved by a sprinkling of silvery white guard hairs. These are thickest in the area of the shoulders, on the posterior portion of the back, and on the top and sides of the head. The underparts, though black, lack the lustrous "finish" so evident on the back and sides. The tip of the tail is pure white in this form also. This is the "silver" fox of commerce, an animal which through selective breeding has become standardized in the fur industry. Nevertheless, the black color is a recessive character, as evidenced by the throw-backs that often make their appearance in otherwise black litters. Without constant vigilance on the part of breeders, the "silver" fox would soon become a rarity again. The Mendelian law cannot be cancelled out by a few generations of selective breeding.

The foxes are the smallest canines native to the United States. Though they look much larger because of their long fur and bushy tail, the average red fox will not outweigh a large house cat. They make up for this lack of size, however, by being exceedingly quick in their movements. They are thus able to catch many of the small mammals which outmaneuver coyotes and wolves. Rabbits are about the largest mammals with which they can cope, but mice, woodrats, pikas, and ground squirrels are all a common part of their diet. In addition, they take many large insects and ground nesting birds and eggs whenever possible. Foxes are not as omnivorous as coyotes, but they relish berries and stone fruits and sometimes raid watermelon patches.

The social life of foxes is most interesting. The family is a closely knit unit which as a rule does not break up until the young are well able to care for themselves. Foxes are monogamous; that is, they normally choose their mates for life. Dens may be in burrows dug in the soil or in deep crevices in the rocks. They are usually in some spot where there is a good view of the surrounding territory. The pups are born rather early in the spring and by early summer will be playing around in the den entrance, although they do not venture to any distance until much later. Should the den be approached while the young are in it, the female often will be very bold in her attempts to lead the intruders away from it. As soon as the young are weaned the male joins his mate in bringing food to them. By early fall, the family is hunting together.

The red fox has been a symbol of sagacity and cunning since long before Aesop. Much of this reputation is well earned, as witness their stubborn withdrawal as civilization surrounds them. Yet sometime one wonders if their wisdom is not overrated. I am reminded of an old female who every year whelped her young in the mouth of a tile drain which drained a marshy piece of ground that had since become dry. The upper end of the tile was buried some 15 feet below the surface of the ground. My friend would watch the area until the pups were about half grown. Then he would block the entrance to the tile with a box trap and catch them as hunger drove them out to the bait. This went on for several years, the old vixen never seeming to learn from bitter experience that her family would be taken away from her.

red fox

Gray wolf
Canis lupus (Latin: dog ... a wolf)

RANGE: Canada and Alaska north to the northern coast of Greenland. In the United States it is found in three widely separated areas in Oregon, Utah and Colorado, and New Mexico and Arizona. It extends south into the tablelands of Mexico.

HABITAT: In the Southwest the wolf, like the coyote, is leaving the plains, which are its chosen habitat, to live in the broken country of the Transition Life Zone.

DESCRIPTION: Doglike in appearance, but larger than a big dog. Carries its short, bushy tail above the horizontal when traveling. The gray wolf is almost unbelievably big. Total length 55 to 67 inches. Tail 12 to 19 inches. Height at shoulders 26 to 28 inches. Weight 70 to 170 pounds. These animals show a tremendous variation in color, but the average individual will appear very much like a big German shepherd dog. From this average, they will vary from the almost white coat found in Alaska to the black phase of the red wolf of Texas. The head of the wolf is distinctive. It has a broad face with a wide but short nose. The straw-yellow eyes have round pupils. The ears are short and round, much more like a dog's than a coyote's. The feet, in keeping with the rest of the body, are large. The front feet have five toes; as is usual with canines, the first toe or "thumb" does not touch the ground. The hind foot has but 4 toes. These animals have a high reproduction rate. Each year the single litter may consist of from 3 to 4 to as many as 12; the average is assumed to be from 6 to 8.

The wolf's association with man is older than recorded history. When man first gained his ascendency over other mammals, the wolf is believed to have been the progenitor of the dog. As man's partner in the chase, it helped him become the one superior animal capable of exterminating it. At the present time, man has come close to doing just that. Only a few of these magnificent wild dogs remain in the United States. Those are concentrated mainly in the Southwest, and some of them undoubtedly have come across the border from Mexico. Before long the species probably will become extinct in this country, but the large numbers remaining in Alaska and Canada should persist for many years.

Much of the public antipathy for wolves comes from literature. Who, as a child, has not thrilled to the danger that surrounded Little Red Riding Hood, and rejoiced at the ultimate end of the arch villain? Long before animated cartooning took over nursery rhymes, children's books were well thumbmarked at the page where the "big bad wolf huffed and puffed and blew the house down." To "keep the wolf from the door" is an expression as full of meaning today as it was in the 15th century when the animal became extinct in England. The wolf has always been a symbol of taking ruthlessly. The genus *lupinus* (Latin: wolf), a beautiful group of plants of the pea family, is so called because early botanists thought it robbed the soil. The "wolf" so often encountered at house parties is included in this class. None of these characterizations gives a good impression, and all are indicative of man's feeling toward the wolf. It is most unfortunate that man so often condemns anything which interferes with his own economic progress. Nature has a place for the wolf, a specialized task for which it is admirably adapted.

In the days before the white man, bison roamed the western plains in great herds which were constantly followed by packs of wolves and coyotes. As long as the bison remained close together they were relatively safe, but woe to the sick or weak that lagged behind. These were quickly pulled down, and after wolves had eaten the choicest portions, the coyotes and vultures moved in for the rest. When the white man exterminated the bison, the wolves' host was gone and they turned to the logical substitute, the white man's cattle. This could have but one result. In the predator control campaign which followed, a wedge was driven through the wolf population of the Southwest, leaving one group isolated in Utah and Colorado and another in southern Arizona and New Mexico. The latter group is actually formed by immigration of wolves from Mexico. It fluctuates in numbers as the animals move back and forth across the border in response to local conditions. During the extermination program, the behavior of the wolf was affected to a considerable extent.

Accounts of early travelers stress the easy familiarity with which the gray wolf accepted their presence. When a wayfarer shot a bison, the wolf sat down within easy range and waited until the choicest cuts had been taken away. It then moved in for its share. Since that time the wolf has become one of the most wary and cunning of our wild creatures. Gifted with a keen intelligence, it has found that only by complete isolation can it escape the methods devised for its destruction. To this end, it has moved from the plains into the more inaccessible places in the mountains. Few will ever see a wolf in the Southwest again, and I consider myself fortunate to have seen this gray ghost of the plains in years long past, and to have heard its deep howl break the silence of a cold winter night.

gray wolf

Coyote
Canis latrans (Latin: dog ... barking)

RANGE: The coyote is common throughout the Southwest.

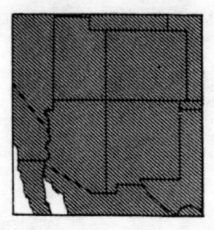

HABITAT: This little wolf, once a creature of the prairies, now is found in all life zones and among many different associations.

DESCRIPTION: Because of their varied associations and wide climatic range, coyotes are of many sizes and colors. In general, they resemble a rather small, lean German shepherd dog with yellowish eyes. A good field mark is the bushy tail which is carried low while the animal is running and seldom is elevated above the horizontal at any time. Average total length 43 to 55 inches. Tail 11 to 16 inches. Color tawny to reddish gray with white or light-colored throat and chest, dark legs and feet. There is usually a dark median line down the back, and the tail also is somewhat darker than the body. Coyotes are lean animals; despite an impression of bulkiness suggested by the long fur, a large coyote seldom weighs more than 30 pounds. The track is much like that of a medium-sized dog; however, the prints of the claws tend to converge toward a center line more than those of the domestic animal. Coyotes are moderately prolific. The average litter contains from 4 to 6 pups, although as many as 11 have been recorded. The best indication that coyotes are in an area is their "singing" during the evening. They will sometimes greet the sunrise, but are infrequently heard during the day.

There probably is more controversy about the status of the coyote in its relationship to other animals than any other North American mammal today. The solution to the argument can be found by taking a 10 minute walk through a bit of the great outdoors. Those living things, plant or animal, which cannot adapt themselves to most changing conditions presented by a slowly dying world must perish. Those which survive do so because they have a mission to fulfill; they must give as well as take from their environment. To me, the unequalled ability of the coyote to withstand the campaigns of man toward its extermination indicates that this animal must be an especially favored child of Nature. Certainly many of the subtle relationships which it maintains with its associations have never been fully explored and others have not been discovered.

In the light of recent studies and with the influence of excellent documentary films in its favor, the coyote's place in Nature is now becoming better known to the public. There seems to be no valid reason why people, who in general like dogs, should express indifference to the fate of this little wolf, which is but a wild dog with what most naturalists agree is a higher degree of native cunning and intelligence than that of the average domestic breed. In general, this attitude seems to stem from unfavorable and usually inaccurate stories circulated by word of mouth. A few hours spent in reading the scientific literature on the coyote will disprove many of these folk tales. For lighter reading try J. Frank Dobie's *The Voice of the Coyote* (Little, Brown & Co., Boston 1949) or *Sierra Outpost* (Duell, Sloan & Pearce, 1941) by Lila Loftberg and David Malcolmson. These delightful accounts present the coyote for what it is—one of the more important creatures in animal society.

coyote

When the first whites pushed their way across the western prairies, the coyote was chiefly a plains animal. Here it lived along the fringes of the huge bison herds, seldom venturing to make its own kills but sharing with the vultures the remnants left from those of the big gray wolves. With small game it was more successful, making heavy inroads upon the rodent and rabbit population. Then, as now, the coyote was also a scavenger and helped rid the plains of the carcasses of larger animals which died of natural causes. When the bison and wolves were practically exterminated, the coyote "took to the hills" and now is as frequently encountered in the higher mountains as anywhere. Farther west in the desert areas the story has been much the same. As civilization has advanced, the coyote has stubbornly retreated into the hills until now its "song" is heard in the highest canyons. The medium size and omnivorous tastes are factors which probably have much to do with its success in this new environment.

About half way between the gray fox and gray wolf in size, the coyote is large enough to subdue the big hares, yet nimble enough to catch the smaller rodents which make up a large part of its animal diet. The rest is supplied by a long list of other small creatures which are less often encountered, including birds, reptiles, and insects. The vegetable portion of its food is no less varied. Berries, stone fruits, cactus fruit, various gourds, some herbs, and even grass are eaten in considerable quantity, depending on the season and availability of meat. Besides this diet of what might be called fresh food, the coyote will usually take carrion. This is the basis for many unfounded accusations against the species. Because scats are sometimes composed almost entirely of the hair of such large mammals as deer, elk and mountain sheep, the coyote is thought to be killing these animals. Actual records of such occurrences are rare; the coyote is not built for such big game. Nature meant this to be the province of the gray wolf. Should such predation by coyotes take place, some other factor undoubtedly would restore the balance before long. Nature's laws are as definite as those of human society and far more sternly enforced.

The family life of these intelligent creatures is interesting in its variations. No two pairs will follow any given pattern. As a rule coyotes, like wolves, will mate for life; but should one be killed, the other will usually seek another partner. Breeding takes place in early spring, followed some 60 to 65 days later by the appearance of a litter of up to 11 pups. The den is usually at the end of a burrow dug in soft soil close to a vantage point which overlooks the surrounding area. More rarely the den is chosen in a crevice among the rocks, and some have been found which are no more than hollows in the shelter of overhanging shrubs. During early life of the pups the male coyote is not allowed to approach them. Later, when they are able to take solid food, he brings his offerings to the neighborhood and the female carries them to the young. Up until the time the pups are able to leave the den, both parents are extremely wary in their approach to the area. They usually come in down wind so as to detect the presence of an intruder. If a human investigates too closely, the pups are moved to a new location at once.

When the young are big enough to emerge from the den, a new phase of their existence begins. At first, they play around the entrance like a group of collie pups, stopping now and then to survey this wonderful new world with wide eyes. Soon the wandering instinct asserts itself, however, and they begin to make short sorties away from the den. This is the time the parents have been anticipating. Now the young can be taken away from an area which becomes more dangerous with every passing day. The family may now hunt as a unit, initiating the young into the coyote way of life, or the mother may scatter the young along the perimeter of her range, bringing food to them as she makes her rounds. In either event, they soon learn to fend for themselves and by the following spring are mature animals.

Unlike his larger relative, the gray wolf, which is a great traveler, the coyote will establish a range and stick to it. In time, he will learn every yard of it and will notice the slightest changes. This is of great importance, not only in evading attempts on his life but also in the matter of filling his stomach. The woodrat, which tonight may be deep within its fortress of rock and branches, will be remembered and called upon again tomorrow when it may be out foraging for pinyon nuts. The cottontail, which reached the brush pile last night, may be intercepted en route tonight.

Several coyotes often share the same range and hunt together. This is especially true of a mated pair which is feeding young. Such a combination is especially efficient in running down such animals as jackrabbits and, more rarely, pronghorns. These creatures tend to run in a circle, and the coyotes alternate in chasing and resting until the animal is exhausted. Then they both close in for the kill. Pronghorn hunting is fraught with danger, however, especially during the time their young are small. These sharp-hoofed animals have been known to pursue and kill coyotes.

It is to be hoped that the relentless persecution of the coyote will soon be a thing of the past. The species has an important place in the ecology of the Southwest, and it cannot be removed without seriously affecting the status of its associates. This is a situation that is deplored by anyone interested in natural history. It is unthinkable that the West should lose this colorful species that is so interwoven with its legends and history.

Wolverine
Gulo luscus (Latin: having to do with the throat ... one eyed; purblind)

RANGE: Canada and the high mountains of California, Utah, Colorado, and possibly New Mexico.

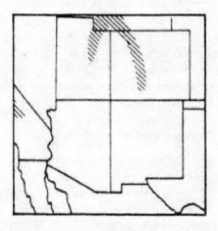

HABITAT: Near timberline in the most remote areas.

DESCRIPTION: A large (20 to 35 pounds), dark-colored animal somewhat resembling a small bear in build. Total length 36 to 41 inches. Tail 7 to 9 inches. In coloration the wolverine shows variation, but with no sharp contrasts. The back is dark brown, shading to a paler color on top of the head. The sides of the body are marked with dull yellowish bands which begin at the shoulders and join near the root of the tail. The underparts are lighter and usually a "blaze" or spot of white decorates the front of the chest. The legs are short and exceptionally powerful, the large feet are armed with long, horn-colored claws. These register rather prominently in the track which otherwise is somewhat like that of a large bobcat. The breeding habits of the wolverine are not well known, but it is assumed the den is located among rocks in talus slopes. The average number of young is thought to be four or less. They are born early in the year.

This mammal, largest of the weasel family, possibly will never be seen by anyone who reads these lines, so scarce has it become in the United States. Yet, because it is such a notorious animal and so little understood, and because it has been recorded in both Utah and Colorado several times, and long suspected to have been a native of New Mexico, it is here included. It would be a shame, indeed, for a layman to see this celebrated creature and not be aware of this unusual good fortune.

The wolverine has been an object of fear and revulsion not only to the white man but to the Indian. It seems to be one of the few mammals which goes out of its way to create destruction and carries a chip on its shoulder toward all other animals which interfere with its desires. It is a creature of mystery, whose life history at this late date we shall probably never fully learn before it becomes extinct.

When the Hudson Bay Company trappers invaded upper North America they found the Objibwa Indians living in a sort of armed truce with the wolverine. They called it "Carcajou," a term said to have been derived from the Algonquin, and accorded it the respect due a malevolent spirit. I have forgotten the Chippewa name for the animal, but I well remember that it was considered a "windigo" or evil spirit. Eskimos coveted its fur for trimming the hoods of their parkas. The long guard hairs protected the face from the bitter air without collecting frost, and the underfur did not collect snow and frost like other furs.

wolverine

The scientific name of the wolverine is interesting. *Gulo*, the Latin term for throat, no doubt has reference to the gluttonous habits of the animal. *Luscus*, also Latin, means one-eyed or, as some authors suggest, blind. This may refer to the small eyes, so deeply set as to be almost invisible at a little distance, or may date back to the first wolverine taken to Europe from Hudson Bay. This specimen was said to have lost one eye, and the name may have been derived from that. At any rate, the normal wolverine is neither one-eyed nor blind.

The wide distribution of the wolverine provides an admirable example of what life zones mean. This same species lives at timberline in the high mountains of desert country and is also found at or near sea level far north of the Arctic Circle. It is well adapted to this environment, with exceptionally thick and heavy fur which does not mat easily with snow. In addition, during the season of greatest snowfall, the edges of the feet and toes grow stiff hairs which, in effect, act as small snowshoes, and enable the animal to travel with less effort.

Food habits of the wolverine are far from selective. Heavy and clumsy in build, it is doubtful if many large game animals fall prey to this awkward hunter. However, it does not hesitate to drive larger predators away from their kills and appropriate them for itself. At such times it eats as much as it can, then hides the rest for future repasts. It will return to the site until the remains are completely devoured, even if they spoil in the meantime. Natural prey includes rodents which it can dig out of burrows, and such ground-nesting birds as it comes across in its travels. It is said to be one of the few successful predators of the porcupine. Thief, predator, and scavenger, the wolverine roams its isolated ranges feared by hunter and hunted alike.

The wolverine is one of the few animals that seems to take malicious delight in harassing human beings. Though robbing of traps can be explained by hunger, theft and destruction of the traps themselves seems to represent deliberate and clever planning. So, too, does the breaking into and entering of isolated cabins with attendant pilferage of their contents. What cannot be eaten is either broken up and defiled or carried away and hidden.

Marten
Martes americana (Latin: a marten ... America)

RANGE: North America from Alaska through the greater part of Canada, thence through northwestern, United States and south into California, Utah, Colorado, and New Mexico.

HABITAT: Usually coniferous forests of the Canadian Life Zone up to the Alpine Zone.

DESCRIPTION: In the trees, this animal is often mistaken for a large squirrel. On closer inspection it will resemble a house cat with a short, bushy tail. Total length 22 to 27 inches. Tail 7 to 9 inches. Weight 2 to 4 pounds. The coloration of the marten is distinctive. The body is a beautiful, soft, yellow-brown, darker on the back, legs and tail. On the chest the color lightens to a pale buff or sometimes a rather distinct orange. The underparts are lighter than the rest of the body. The fur is extremely fine and thick. It is distinctive in being almost entirely underfur, there being very few guard hairs. The body is extremely graceful with relatively long legs and small feet. The head is small with features somewhat resembling those of the weasel. The ears are large for a member of the weasel family and lend an alert appearance to the face. This alertness is further borne out by the lively movements of this animal, which is the most active of any in that group.

The marten, often called "pine marten," is one of the most solitary animals of a group whose members habitually travel alone. Perhaps this is because in this family of predators each species is fully able to overpower any resistance put up by its accustomed prey, individually and not through force in numbers. Perhaps, too, it is because the entire group is made up of voracious eaters which, if they ran in packs, could not encounter enough prey to adequately feed them all. Finally, this clan has several species which instinctively kill far in excess of normal needs. This is a practice which, almost without exception, is confined to those members of the weasel family which prey on rodents. It is evidently one of Nature's methods of controlling the rodent population. To operate at highest efficiency these killers should hunt alone. These factors all apply in some degree to the marten. As a consequence, although there may be many in an area, the marten is usually found alone except for a brief time during the breeding season or in the case of a female with young. The male evidently has no part in bringing up the family.

The marten has always been more or less plentiful throughout its range, and there is no reason to believe that it will not continue to be seen by alert observers for many years to come. Its chosen habitat is among the evergreens near timberline. This is also an area of rock slides, and the marten loves to hunt the small rodents which make their homes there. Indeed, it divides its time between the two environments, hunting in the talus slopes during summer months, and taking to the trees in winter when rock slides are buried deep beneath the snow. It is an extremely hardy creature which holes up in an abandoned squirrel or woodpecker nest only during the short periods of storm, when hunting would be useless. As might be expected, its summer and winter diets vary widely. Both, however, have as their basic item the spruce squirrel, the important host of the marten, and like it a hardy creature that is abroad throughout the year.

There is considerable variety in the summer diet. On and in the ground there is available an amazing number of species which are denied to the marten during the winter, some because of protection afforded them by the deep snowdrifts and others because they hibernate. Among these are pikas, ground squirrels, woodrats, chipmunks, and many species of mice. In summer, the marten also takes eggs and young of ground-nesting birds. In the trees are found other nests, not excepting those of the woodpecker, into which the marten inserts its forepaw and comes out not only with young birds, but often the adult as well. Martens are known to eat quantities of the larger insects and, since they are fond of fruits and berries when raised in captivity, there is little doubt that they indulge in these delicacies in the wild.

Winter diet consists of the spruce squirrel, augmented by such other small creatures as may be abroad during cold weather. Though it would seem that the marten might suffer from the curtailment of its lavish summer menu, the opposite is the case. They remain fat and healthy under weather conditions that would seriously hamper most other predators. To a large extent, this ability to survive is due to the untiring perseverance and great skill with which they hunt. In addition, few creatures have been endowed with so many adaptations with which to withstand the long, cold winter.

marten

It will be apparent, even to the casual observer, that the marten is most precisely evolved to meet the frigid conditions imposed by its boreal habitat. The long, fine-haired winter coat is extremely warm and does not mat with snow or frost. With such an insulated covering any hollow log or woodpecker's nest will do as a resting place. Snow is the least of the marten's troubles; not only does it stay warm among the drifts, but travels across them with ease on its "built-in" hair snowshoes, which also keep the toe pads warm. The midwinter track of a marten is rather confusing, as it shows no definite toe marks, but is a blurry outline in soft snow, and on harder snow scarcely registers at all. However, if it is remembered that this animal travels much like a weasel, that is, it jumps instead of walking, the larger prints will serve to identify it as a marten.

Interesting as the physical adaptations of the marten may be, the response of its life history to the pressures of a long winter are no less fascinating. As has been stressed, the marten is a solitary and more or less nomadic animal. Apparently the only time of the year that is favorable for breeding is during the summer, as this is the only time when adults of the two sexes are commonly found together. This starts a reproductive cycle which, while not too uncommon, is unusual enough to excite one's interest. For the following information, I am indebted to James Campbell of Hope, Idaho, who live-trapped and raised many of these interesting animals years ago when knowledge concerning them was relatively meager.

Box traps were used to take the marten during the middle of the winter, when snow lay from 15 to 25 feet deep along the trap lines. This was at an elevation of up to 6,500 feet in the panhandle of northern Idaho. As a sprung trap was approached, the outraged captive could be heard growling its resentment and struggling to escape. A flour sack would be placed around the entrance and the door opened. The marten, apparently mistaking the white glare for snow, invariably would leap out into the sack. Great care was necessary at this point, for the marten was usually wet with perspiration from its struggles within the box trap, and if allowed to chill would quickly die from exposure. The sack was placed within several others and the bundle placed in a pack-sack and carried down the mountain, where the marten was cooled gradually in the house, then put in the outdoor pens. Here they soon became so tame that they would readily accept food from the hand, never becoming treacherous like their unpredictable cousins, the mink. They loved fruit and berries, and were especially fond of chocolate candy.

Early in the venture, it was observed that winter-caught females were giving birth to young in April. Further observation revealed that breeding took place from the early part of July into late August, but that no matter when breeding was accomplished the young would be born in April. The first signs of pregnancy, however, would not be apparent until about 50 days before birth of the young. This indicates that, like most of the hibernating bats, breeding takes place in one season, but the fertilized ova remain quiescent and do not begin to develop until conditions are propitious for the birth of the young. This also insures arrival of the little ones quite early in the season, so that they may enter the following winter fully grown. The number of young varies from three to five, usually the smaller number.

No description of the marten would be complete without mention of its tremendous vitality. In trees it is superior to the squirrel, especially in long, arching leaps, which it makes from one lofty perch to another. In winter time it will often leap from the trees into soft snowdrifts, seemingly for the sheer thrill of the sport. It is not uncommon for martens to burrow through snowdrifts for some distance apparently in search of rodents. I have found that a marten, startled in the forest, is not usually too afraid of its arch enemy, man. At first it will run away but, if pursued too hotly, will come to bay on a low limb and put on a great display of hissing and growling while baring its sharp, white teeth. It is not improbable that if it were pressed further it might attack its tormenter.

River otter
Lutra canadensis (Latin: otter ... of Canada)

RANGE: Most of North America south to central Arizona and New Mexico in the Southwest, and south to the Gulf of Mexico in the east.

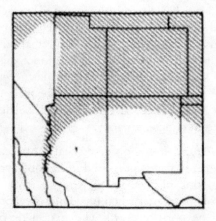

HABITAT: Along and in fresh water streams and lakes.

DESCRIPTION: A short-legged, stream-lined creature with a thick tapered tail, usually seen in the water. Total length 3 to 4 feet. Tail 12 to 17 inches. Weight up to 20 pounds. Color mostly a rich, dark brown with a silvery sheen on the underparts. The throat and chest are lighter than the rest of the body. The otter is well adapted to aquatic life, having a long, round body and short, muscular legs. All four feet are webbed. The head is long and round, with short ears. Long, stiff whiskers stand out near the rather thick nose. The tail is thick at the base, and the body literally tapers off into the tail, increasing the general "torpedo" effect.

river otter

The otter, never plentiful in the Southwest, has become extremely rare in recent years. This is due in large part to its highly specialized habits, coupled with an inability to compete with man in the use of the few fresh water streams and lakes in the desert mountains. Yet, it has been recorded often enough in the past decade to warrant the hope that with careful management and complete protection it might increase in numbers. This is much to be desired because the otter is unique in several respects among our native mammals. This mild-mannered member of the weasel family lacks many of the fierce and blood-thirsty habits of its more ferocious relatives. It is, instead, gentle, even playful.

Outstanding among these characteristics is the otter's habit of building slides. These are probably nothing more or less than a refinement of the way otters travel through the tules and slippery mud flats, in which they spend much of their time hunting crayfish and small amphibians. The remarkable thing about the slides is that they seem to be built for one specific purpose, that of sport, an activity which ordinarily is one of the least important to most mammals. In soft or muddy places, even in soft snow, the otter slides along on its chest with head held high and forelegs trailing alongside the body. Motive power is furnished by thrusts of the hind legs. Excessive wear on the underparts is reduced by many coarse, close-set overhairs which seem to have been developed for this very purpose. The slide itself is only a narrow groove, 12 to 20 inches wide, that is worn down a steep bank to the water's edge. The wet bodies of the otters make it smooth and slippery, and soon they are able to shoot down it with only an occasional helping kick of the hind feet. This fascinating game may go on for hours on end. The descent often is followed by a general rough and tumble in the "swimming hole." There the action is almost too fast for the eye to follow, because few mammals can match the otter for grace and speed in the water.

Aquatic as the otter is, it does not care to be always wet, and this leads to another curious institution in its way of life. Near the slide, and usually at several other places along the waterway which is frequented by a family of these delightful creatures, will be found areas several feet in diameter, located among dry tules or in tall grass, where the animals roll and thus dry themselves. These seem also to be community news centers, because usually near such areas are found the scent "posts," where otters deposit scent from the glands common to all members of the weasel family. In otters these glands do not secrete the high-potency perfume produced by those of skunks and minks. Nevertheless, it is sufficiently "loud" to be identified with the otter.

The dens present great contrast in location and type. They are usually situated near water, but one was found more than half a mile from the nearest stream. On the other hand, an otter will often take over the abandoned burrow of a bank beaver, and access to this abode must be by an underwater entrance. In many instances, the den is merely a nest in a thick clump of tules completely surrounded by water.

The two to four young are born in early spring. At birth they are blind, toothless, and amazingly helpless in comparison with their development 6 weeks later. At this age they begin to leave the den, and before long are quite at home in the water. Though the male may be in the neighborhood, the female will not allow him near the young until they are half grown. At this time, the family will begin to live together until the young are fully able to make their own way.

Otters are cosmopolitan in their tastes; being carnivores, they will prey on many species. Fish is their preferred food, and in most cases they capture rough fish species, these as a rule being slow and easy to catch. They are fully capable of catching trout, however, should other supplies fail. Otters in captivity do not thrive on fish alone, so evidently the great numbers of other small animals upon which they prey must be necessary adjuncts to their diet. These include crayfish, frogs, several species of small mammals, and such birds and eggs as may be available.

The presence of otters in an area is not difficult to detect. A slide, "rolling place," or characteristic web-toed track are all sure indications that this interesting animal is a neighbor. Cultivate its acquaintance if you can. The otter is diurnal as well as nocturnal, and should you be so fortunate as to see this happy animal coast down his slippery slide, I am sure you will get as big a thrill from it as he does.

Mink
Mustela vison (Latin: weasel ... forceful, powerful)

RANGE: The range of the mink is strikingly similar to that of the otter, that is, it embraces most of northern North America, extending southward into southwestern United States in the west, and to the Gulf of Mexico in the east.

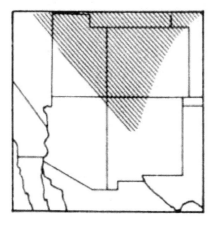

HABITAT: This semi-aquatic animal seldom is found far from fresh water streams or ponds.

DESCRIPTION: The mink is about as long as an average house cat, but is much more streamlined in appearance. Total length for males 20 to 26 inches. Tail 7 to 9 inches. Weight up to 2¼ pounds. Females will average almost one-third smaller. Color is dark brown over most of the body, shading to lighter brown on the sides and darkening along the tail to a black tip. There are usually a few irregular white spots on chest and belly. The body is long, and round, tapering into the long, round neck. The head is small with rather a triangular face, small ears, and dark, beady eyes. The legs are short and, as would be expected on an aquatic animal, the feet are webbed, but in this case only the bases of the toes are joined by the webs. The underfur is thick and fine, the guard hairs coarse and conspicuously shiny. Mink will bear as many as 10 young, but the average is around 5. Dens usually are in a burrow, which may or may not have an underwater entrance.

The presence of mink in any given area is usually quite easily determined by scouting sand bars and mud flats along the water's edge. The tracks are quite distinctive, especially in softer mud, because here the animal spreads its toes to keep from sinking, and in places the outlines of the partially webbed toes become clearly apparent. In most cases if tracks are at all discernible, marks of the claws are conspicuous. The occurrence of mink away from water can not be considered normal, because this creature ranks second only to the otter, among southwestern carnivores, in its preference for an aquatic life. Exceptions do occur, however; mink have been encountered crossing mountain ranges where they might be many miles from the closest watercourse. It is thought that these infrequent cases may be migrations from unfavorable areas, or that such a trip may be undertaken in search of a mate.

Much of the mink's dependence on water stems from its diet. Some of its preferred foods are fish, crayfish, and frogs, none of which are more adept in the water than the mink. Other food items, taken whenever circumstances permit, are birds and eggs and rodents. It is interesting to note that the muskrat is no match for the agile mink, and that one of these fierce carnivores moving into an area has resulted in the extermination of a whole colony of muskrats. Cottontails, too, are unable to cope with the tactics of the mink, although their reproductive proclivities usually keep their numbers well ahead of such inroads. Even with this wide variety of prey and its expertness at hunting, the mink is so voracious that in some areas it has been estimated 100 acres are only enough to support one adult. The continual hunt for food may be the motivation for another interesting habit of the mink which is seldom found among other carnivores.

Many beasts of prey will hide or bury a kill and come back to it later for several more meals. In fact the wolverine, one of the mink's close relatives, will do this. However, the mink actually collects a considerable store of food during periods of good hunting and caches it away against time of need. Caches will often consist of larger animals, such as muskrats and ducks, laid neatly away under an overhanging bank. Since these stores are highly perishable, this is mostly a cold weather practice. The mink is not normally a carrion eater.

A characteristic of the weasel family is the occurrence of anal glands which secrete a liquid having a powerful odor. The skunks are best known in this respect. In my opinion the mink and weasel both release an odor which, by comparison, makes the skunk's "almost nice." The one saving grace in their case is that the odor soon evaporates, while that released by the skunks retains its strength for a long time, and regains much of the original potency with every rain. Like the skunks, these animals use the disagreeable odor as a defensive weapon. It no doubt has other uses too, such as identifying the individual and its territory to other animals of the same species.

Considering the weasel family as a group, it becomes apparent that here is a rather large number of species, all closely related, yet having widely divergent habits. For instance, the marten is as much at home in trees as is the squirrel; the otter can catch fish with ease; and the badger is able to dig better than even the ground squirrels and spends much of its life underground. In the same way, the group varies widely in temperament. At one end of the scale stands the wolverine, surly and defiant; at the other are the marten and otter, playful and even affectionate. The mink might be classified as nervous and irritable. There seems in its temperament to be an actual blood lust. When the mood is upon it, it will continue to kill even when a human is close by. I have seen a mink continue to slaughter a flock of ducks even as I was attempting to drive it away. A mink cornered is a creature to reckon with; there are few animals its size that are so courageous.

mink

As might be suspected, such wildly fierce creatures make poor parents. The females sometimes desert the young while they are still too small to make their own way. Yet this, after all, is but a human criticism. Who is to condemn an animal which Nature has allowed to exist under conditions that would have eliminated a more amicable species?

Short-tailed weasel (ermine)
Mustela erminea (Latin: weasel ... from the fur ermine)

RANGE: From northern Greenland south to northern United States with one extension south into Utah, Colorado, and New Mexico. To be expected in northern Arizona.

HABITAT: Generally found in forests of the Transition Life Zone and higher. It will often be found in the Arctic Zone.

DESCRIPTION: A tiny predator with long body and short legs. Total length from 7 to 13 inches. Tail 2 to 4 inches. Weight 1½ to 3⅔ ounces. This wide range in statistics is from comparing the smallest females with the largest males. Males consistently average from one-fifth to one-fourth larger than females. Summer color is dark brown with white underparts and feet. There is a white line down the insides of the hind legs connecting the white of the feet with that of the belly. The tip of the tail is black. Winter coat is all white with the exception of the black tail tip. The body is long and supple, legs are short, the neck long and round. The head is small with rather large, bulging dark eyes. The ears are large for a creature of this size. Breeding dens are usually in the ground under large rocks or among the roots under a tree. Average number of young is thought to be about four.

I have a special affection for this tiny predator which, because of its fearlessness, has given me many a glimpse into its private life which would not have been possible in the case of a larger or more timid creature. Let no one underestimate the courage of this small mustelid which, if left alone, will continue its normal activities even under the close scrutiny of an observer, but if molested will often turn on its tormenter with a fury matched by few large animals. It shares these characteristics with two other relatives of the United States: the longtailed weasel (*Mustela frenata*), which is also found in the Southwest, and the least weasel (*Mustela rixosa*), which inhabits part of the northern United States, Canada, and Alaska. The short-tailed weasel will not be mistaken for either of the other species, since the least weasel has no black tip on the tail and the long-tailed weasel has a tail about one-third of its body length. The tail of the short-tailed weasel is only about one-fourth of its body length, and this species is considerably smaller than the long-tailed weasel.

Short-tailed weasels are the smallest carnivores in the Southwest. In fact, except for the least weasel, they are the smallest on the North American Continent. Despite its size, *Mustela erminea* is so hardy it ranges to the northernmost point of land in the Northern Hemisphere. This, the north coast of Greenland, is but a few degrees from the North Pole. The European form, not specifically distinct from ours, is equally hardy. It, too, inhabits not only the more temperate zones, but penetrates far north of the Arctic Circle wherever land is found. In our Southwest they are sometimes encountered at low elevations but more often in the higher mountains. Here they go through the winter change of color, but not so regularly nor so completely as in the far north.

The term "ermine" refers to this animal's fur in the winter pelage. This is the royal ermine, reserved in days past for the use of the aristocracy. At its best this fur is a spotless white, except for the sharply contrasting black tail tip. In heraldry the pure white had symbolic significance, but to the weasel it has more mundane uses. These are as camouflage, both in pursuing prey and in avoiding attacks of enemies. In the far north this seasonal change of garb is mandatory and complete, but in the mild (by comparison) climate of our southwestern mountains the situation is somewhat altered. Here the creature can descend to lower elevations as winter comes on and, if it wishes, evade most of the severe weather. Under conditions which to some extent are left to its own choice, the degree of color change varies greatly. In snowy areas on higher peaks it will change to true ermine; lower down it probably will turn to a light yellow, and below snowline the animal will retain the same brown above and white below that it wears all summer.

short-tailed weasel

Like most other members of the weasel family, these small mustelids are admirably adapted to do their part in Nature. Their size permits them to enter the homes of all but the very smallest rodents. Their strength and suppleness combined with ferocity enables them to subdue animals several times as large. Surprisingly enough, though well able to climb, they do not eat many birds. Most of their prey is rodents. Small mice seem to be preferred, though chipmunks, ground squirrels, and woodrats also are taken. Pikas and small rabbits fall prey to these mighty mites, and there are many recorded cases of snakes being killed by them. Like the mink, short-tailed weasels will gather a cache of food when hunting is good. For their size they have a tremendous appetite; it has been estimated that one will eat half of its own weight in food every 24 hours. From this it will be seen that they can live only in an area where rodents are plentiful, and that they play a large part in keeping these creatures under control.

I have been privileged to see this weasel many times and under varying circumstances. In all of these encounters it has seemed evident that at first the animal accepts the intrusion of man not so much as an enemy, but rather as a competitor. Under these condition it will continue its activities and pay very little attention to the intruder. However, should any hostile action be taken against it, the weasel will make its escape, if possible. If cornered it will savagely defend itself, and as a last resort spray its attacker with the foul-smelling contents of the anal gland. Not so long lasting as the skunk's perfume, this odorous mist is nearly as effective while it lasts. How much better to stand aside and watch the little predator go about its work!

If you are fortunate enough to be in an area where a hay meadow is being irrigated, you will see the meadow voles (meadow mice) being flooded out of their homes. A careful watch may reveal one or more short-tailed weasels taking their toll of these hapless refugees. You may even find a cache laid away during this period of good hunting. Neither pity the voles nor scorn the weasel; both are only fulfilling their destinies in an ages-old plan.

Spotted skunk
Spilogale gracilis (Greek: spilos, spot and gale, weasel ... gracilis, Latin: slender)

RANGE: This species, together with several subspecies, is the common spotted skunk of the Southwest. It has a "spotty" distribution over the whole of the four-State area with which this book is concerned.

HABITAT: Common in most situations which offer suitable environment from near sea level, to an elevation of approximately 8,000 feet. Seldom encountered above timberline. These skunks normally live in burrows in the ground, but are not averse to taking up residence under buildings or in the walls or attics of frame houses.

DESCRIPTION: A small, nocturnal, black and white animal about the size of an average grey tree squirrel. Total length about 16 inches, of which 6 inches is taken up by the tail. One description of the color pattern would be to call it marbled. The head usually has a prominent white spot between the eyes, with several smaller spots on the sides of the face. The forequarters are marked with four lateral, irregular white stripes which reach to mid body. The rump is variously blotched with white. Tail very bushy and about half white and half black. Eyes dark in color, ears small. Feet small but plantigrade as in the larger species of skunks. Young number three to six, born in early summer.

Although this little animal has a slight heaviness of the hind quarters, reminiscent of the larger skunks, it is indeed, as both generic and specific names suggest, much more like a weasel. This impression is heightened by its quick movements and a bright-eyed attention to details which its larger relatives would hardly notice. It lacks the wild and fierce disposition of the weasels however, and becomes a charming and confiding nocturnal visitor if properly encouraged. Remember though that this acquaintance can be no more than an armed truce, and that should the articles of Formal Conduct be violated it can be terminated at a moment's notice.

Probably no nocturnal mammal in the Southwest is more likely to be encountered than this little skunk. How many of my readers can recall drifting up from an uneasy sleep to the sibilant whisper of, "there's something in the tent." While eyes strain to pierce the darkness, faint patterings on the floor and urgent scratching at the grub box indicate that there is indeed "something in the tent." Turning over with the utmost care, while the joints of the cot loudly complain, the flashlight under the pillow is finally extricated. Surely the creature has been frightened away, but no, the rattlings continue—in the dishes now. The brilliant white beam stabs in that direction. Red eyes stare back, interested perhaps, but unafraid. The rounded ball of black and white fluff waits motionless to see if any harm is intended. When none is offered, his highness makes his way to the door and ambles away into the enveloping darkness. In the morning tiny squirrel-like tracks in the dust show that *Spilogale* has paid a nocturnal call. These, and perhaps the contents missing from the butter and bacon grease containers, because this little animal dearly loves animal fats. These are the foods which attract these animals to camp sites in such numbers that they frequently become a nuisance.

In the wild, spotted skunks live largely on insects. These are taken not only in the adult form but also in great numbers in the larval stage, as is shown by the well-winnowed debris under clumps of cactus and around the bases of shrubs and trees. In these searches for insects small prey of other kinds is captured as circumstances permit. Worms and scorpions as well as small rodents are not refused. More rarely a ground-nesting bird may be disturbed and the eggs or young taken. In rural communities hen roosts are sometimes raided too but in the main the spotted skunk should be considered beneficial, with control of grasshoppers and beetles it's chief function.

Like most predators, this member of the weasel family has few natural enemies. This is not surprising; few animals willingly take a chance on attacking this doughty little warrior, which sometimes does a handstand the better to spray it's enemies. These tactics avail nothing against the steely monsters that rush up and down our highways in the dead of night. In the space of 50 years the automobile has developed into the most successful enemy of the spotted skunk. Yet even in death on the highway the skunk has it's revenge. Few will pass the spot for many a day without paying unwilling tribute to this malodorous legacy.

Striped skunk
Mephitis mephitis (Latin: a pestilential exhalation)

RANGE: The southern half of Canada, the whole of the United States, and the northern half of Mexico.

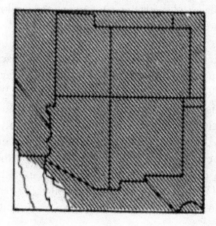

HABITAT: All life zones up to timberline in places which have a sufficient food supply and proper cover.

DESCRIPTION: This is the "wood kitty," approached with due respect by all but the most naive. About the size of a house cat. Total length 22 to 30 inches. Tail 8 to 15 inches. Weight 6 to 10 pounds. Body color is black, with black tail except for the tip, which is commonly white. There are usually two white stripes on the back joining in a "V" at the back of the head and a white stripe down the front of the face. The head is small with a rather pointed nose, small black eyes, and small ears. Front legs are short, and the small feet are tipped with stout claws. Hind legs are longer and appreciably more of the large hind feet touch the ground. The tail is quite long and extremely bushy. It is carried in a downward curve when traveling; if its owner is startled or angry, it is held straight up with the hairs flared out. Dens of the striped skunk are usually in an underground burrow, but dens in hollow logs have been recorded. The usual number of young average from four to six. The family remains together for the greater part of a year before the young leave to make their own way.

There are four species of skunks in the Southwest, but the observer in the higher country will see only two. These are the striped and the spotted. They are distinguished by two characteristics: first, the striped skunk is easily double the size of the spotted skunk; and, second, the spotted has a pattern of broken stripes and spots of white, whereas the larger animal has definitely long, continuous white stripes along sides or back. Both species have the same method of defense, but the odor of the smaller skunk is said to be somewhat less pungent and dissipates sooner than that of the striped. To the recipient of either barrage this has the same consolation as if he were given a choice between being hit by the H bomb or the A bomb. In the event of a direct strike it makes little difference.

Should the reader be involved in an encounter with one of these malodorous creatures, there are many remedies prescribed but few giving any great measure of relief. If the skin is washed with a weak solution of acid such as lemon or tomato juice and then scrubbed thoroughly with soap and water, much of the odor will disappear. Clothes can be given the same treatment, but usually it is cheaper and easier to burn them and charge the cost to experience. Grandpa said to bury scented clothes in damp earth. Perhaps in time this will do the trick; I contend they are better left there.

So much misinformation exists about the skunk's defensive mechanism and the manner in which it is employed that brief explanation may not be amiss. The scent is a fluid stored in two glands located near the base of the tail. These glands are embedded in a mass of contractile muscle, and each has a duct which connects with a tiny spray nozzle that can be protruded from the anus. When danger threatens the tail is lifted, the nozzles aimed at the enemy, and the contraction of the muscles around the glands forces out a spray of fine droplets which may carry as far as 15 feet. The result is usually effective and lasting. Contrary to popular belief, the odor is distressing to the skunk as well as to its enemy. The tail is kept out of the way if possible, since its plumey depths would hold the scent for a long time.

striped skunk

Skunks of different species will use this defensive weapon against each other. Whether individuals of the same species use it in their fights together is not known. In situations involving humans the skunk will try to bluff the enemy if possible. This consists of stamping the front feet, of short runs at the intruder, and finally of hoisting the tail and aiming the "guns." If a skunk is approached deliberately and if quick movements are avoided, it is surprising what liberties may be taken before it will resort to scent. On the other hand, should it be taken by surprise or should it be physically hurt, retaliation is swift and certain. In all cases where skunks are encountered at close range, remember that this little animal is one of the most independent creatures on earth, that this nonchalance stems from a supreme confidence in its defensive powers, and that if left alone or at least treated with consideration it will go on its way as soon as possible.

This independent attitude inherent in all skunks probably has much to do with the happy-go-lucky life that the young family lives. About midsummer when the young are able to leave the burrow, the mother often will take them for a stroll early in the afternoon. As she walks, oblivious to danger, the young play along behind her, sometimes a ball of struggling little bodies with now and then a fluffy tail breaking free and again all at odds in a mock show of ferocity with front feet stamping and flared tails held aloft. When the patient mother finds a tidbit on the trail, there is a concerted rush for the prize, which is seldom won without a struggle. All of this is excellent practice against the time when they will be on their own. It is during this early age that the young first learn to catch insects, items of great importance in skunk diet. Later frogs and small mammals will also be preyed upon.

The striped skunk is generally considered a hibernating animal. This is not strictly true for, while it may remain inactive in its den for weeks at a time, the body processes do not slow down to the extent common in true hibernation. The skunk does lay on a considerable amount of fat each fall in preparation for this period of winter when food is scarce. Actual retirement to a den for even a few days is rare in the Southwest, however. The mild climate makes this unnecessary, except in the highest part of their habitat.

Black bear
Euarctos americanus (Latin: a bear ... of America)

RANGE: At present the range of the black bear in the United States is confined to a narrow strip adjacent to both the Atlantic and Pacific coasts, a few of the southeastern States, a narrow band in the Great Lakes area, and the Rocky Mountain chain.

HABITAT: In the Southwest, the higher mountains mostly in the Transition Life Zone and above.

DESCRIPTION: The black bear needs little description because through pictures and reputation it has become well known to almost everyone. It averages 5 to 6 feet in total length with a tail so short as to be inconsequential. Height at shoulders is 2 to 3 feet. Weight 200 to 400 pounds. Color varies in the Southwest from deep, shining black through brown to light cinnamon. In all color phases the nose is brown almost back to the eyes and there is usually a white "blaze" on the chest. The legs are short and muscular. The feet are plantigrade, that is, the bear steps on the whole foot, not just the toes. There are stout claws on all four feet. The head proper is rather round, the muzzle long and pointed. Ears are relatively small, as are the dark eyes. The young number from one to four, with twins being very common. They are born while the mother is still in winter quarters. When the weather moderates to a point where she can leave, the cubs are large enough to follow her.

Bears are probably the most popular of our wild creatures to those who visit the National Park Service areas. Why this should be is hard to say. Perhaps it dates back to the nursery tale of the three bears, familiar to all of us from the time we were able to walk. Perhaps too it stems from the easy familiarity with which these roadside bandits hail the tourist in hopes of a handout. At any rate, these seemingly friendly clowns have become endeared to the hearts of the American public. This is regrettable because actually in the Park Service areas these big carnivores are the most dangerous of all animals. Native intelligence indicates to the bears that food may be had merely by standing up alongside the road when a car stops. More complicated routines are soon learned to wheedle bigger and better handouts. At this professional level, a substantial reward is expected when Bruin has "sung for his supper," and should none be forthcoming, trouble is apt to ensue. This is but a minor annoyance to a bear, however, when compared with some of the indignities dealt out to these big creatures by a thoughtless public. It must be said in all fairness that anyone who teases a bear deserves whatever is handed out in return. It is unfortunate that retribution may be in the form of serious injury or even death. Though this applies mainly to the half-tame bears which roam along the highways in our National Parks, it is only common sense to avoid incidents with any bear wherever encountered. This is especially true of an old female with cubs, a combination well nigh irresistible to the average vacationer with camera.

In more remote areas where bears have not had contact with man, they are wary to the point of timidity. Gifted with a keen sense of hearing and smell which makes up for their poor eyesight, they are difficult to approach. Like most animals, they instinctively know that by "freezing" they can in most cases escape being seen. The sunburned coat of the brown phase of the black bear is especially hard to spot in the underbrush. However, with patience and the aid of binoculars, it should not be too difficult to get a glimpse into the private life of these engaging creatures.

black bear

Though bears, because of their dentition, are classed as carnivores, they might more accurately be termed omnivores. It is a matter of record that the black bear will eat almost anything, either animal or vegetable. Nevertheless, its appetite is prodigious and demands little variety, if but a few kind of foods are available. Its status as a predator is somewhat confused. Technically speaking, since the black bear preys on ground squirrels, mice and other small rodents it should be classed as a predator. It will also take young deer and elk whenever it can, but these opportunities come rarely. Actually this bear has little direct influence on its mammal neighbors. As a scavenger it has considerable value in cleaning up the remains of kills made by other predators.

Some of the small animals eaten are in almost amusing contrast with the huge size of their enemy. For instance, ants are eagerly lapped up by most bears, and they will literally tear old logs apart to get at these toothsome morsels. Grubs are another small item which may be found around fallen logs and under stones. Bears are extremely fond of honey and will go to great lengths to get at this delicacy, which they eat comb, bees, and all. Another food item which seems unusual is fish. At spawning time a bear will wade out into a stream and either snag a passing fish on its long claws or flip it out on the bank where it is more easily subdued. Finally, their natural animal diet is greatly augmented in most Park Service areas by the scraps and bones which they pick up on the garbage heaps. They can become a great nuisance in the camping areas where, under cover of darkness, their ingenuity and great strength enable them to steal many a ham and side of bacon.

Wide as this variety of animal food seems, it cannot equal the cosmopolitan tastes of these bears in a vegetable diet. Roots and bulbs of many species are dug up. Grass and browse are eaten during several seasons of the year; even pine needles are recorded as having been eaten. The liking of bears for berries of all kinds is well known. *Arctostaphylos*, the generic name of the manzanitas, translated from the Greek means "bear grape." Pinyon nuts, acorns, chokecherries, and other stone fruits all are gathered in season. These heavy animals often damage trees severely in their search for fruit. On the garbage heaps, watermelon rinds and seeds, peelings of all kinds, leafy vegetables, and corn cobs add to the fare. All tin cans are licked clean, and in many cases greasy paper and cellophane wrappings are eaten.

The yearly cycle of a bear's life is a study in contrasts. Much of the warm part of the year is spent in search for food with which to build up a store of fat so that the winter may be spent in inactivity. Bears hibernate or, more properly, retire for several months of the winter. They do not fall into the deep sleep indulged by some of the rodents. Theirs is an uneasy sleep broken by periods of lethargy when they are awake but avoid any activity. By these means they conserve enough of their thick layer of fat to live out the cold weather and emerge in early spring with a considerable reserve.

Hibernation takes place in late autumn, usually after the first light snows. Evidently the animals have a den already located, for when they feel the urge to retire they strike out across country to it. The same winter quarters often will be used by one individual for several seasons. Dens are chosen in a variety of locations. They may be in old hollow logs or in the bases of fire-gutted trees. Some are in crevices among huge boulders, others in caves. The main concern seems to be to find a place sheltered from the wind and snow. If the floor happens to be covered with chips or leaves, so much the better. It usually is, either from air currents which bring in falling leaves or through the labors of woodrats which deposit much litter in such places. The bears curl up on the floor, and after the first heavy snow there is nothing to mark the spot. In the case of a small den, such as a cavity in the base of a tree, an airhole may form in the drift from the warmth of the animal's respiration.

The cubs are born in late winter. From one to four in number, they are incredibly small at birth. They develop rather slowly and at the time the family emerges from the den are approximately 18 inches long. The cubs may all be one color or some may be brown and some black. The male bear has no part in raising the family; indeed, he is driven from the scene by the irate mother, should he approach too closely. She has all the responsibility for raising the family, and a busy time is assured with such mischievous, carefree youngsters.

One of the first lessons learned by young cubs is that of obedience. The mother insists on compliance with her every command, and enforces her authority with a heavy paw. It is fortunate the cubs are sturdily built, for some of the slaps they receive in the course of an average day's instruction would kill a less durable animal. The first haven of refuge when danger threatens is in the trees. A special command note and a slap or two sends them hustling. Now the cubs are out of the way and the decks cleared for action, so to speak. The cubs will remain in the trees until the mother lets them know they may come down. This is not a time of boredom for the youngsters, however. Expert climbers, they carry on the same games and rough play indulged on the ground, with never a fall. Their confidence in the trees is amazing. It is not unusual to see a cub sound asleep on the end of a 20-foot branch that is bending down with its weight and swaying in the wind. As the months go on the cubs begin to lose their juvenile ways. By autumn, they have put on enough fat to last the winter. They usually hibernate with the mother, since they remain with her for well over a year. During the following summer they are well able to take care of themselves, and the mother deserts them.

It is normal, rather than unusual, among black bears to breed only every other year. The youngsters usually do not breed until about three years old.

No account of this bear would be complete without mention of the so-called "bear trees." These are trees situated at the crossroads, that is, near the intersections of bear trails or otherwise prominently located. When a bear encounters one, it stands up and scratches at the bark with its front claws as high as it can reach. Sometimes it will also bite at the bark. Bears have been observed rubbing the sides of their jaws against the bark. Whether this is a way of leaving their scent is not known. It is thought this may be a way of communication with others of the species, but this has not been definitely proven. Many of the trees chosen for this purpose in mountains of the Southwest have been aspens. The heavy black furrows left in the white bark will persist until the death of the tree. Often they are the only evidence that bears have ever been in the locality.

Another custom which will be observed very early in one's experience with bears is the scratching that goes on. It may be due in part to the presence of ectoparasites, but the bear takes such an obvious satisfaction in scratching that, one feels, this must be only incidental. Trees, posts, rocks, and claws are all employed for this purpose. Some of the smaller trees often suffer severe damage from the treatment accorded them.

My cautious attitude toward bears is a result of early experiences with them, ranging from humorous to tragic, and probably best typified by an incident which took place near Yellowstone Park in the late 1920s. I was on my first trip into the Rockies at the time and hired out on a construction job at an isolated dude ranch. Horses were being used, and their supplies, including a considerable store of oats, were kept in a large tent adjacent to that in which some of the employees slept. On the previous night a bear had gained access to the supply tent, torn open a number of oat sacks, and wasted more of the grain than it had eaten. The foreman, an old-time packer in the Park, vowed vengeance on the bear. That night when he went to bed he leaned a small, double-bitted axe against the entrance to the tent. During the night I awoke as the foreman went out the entrance in his underwear. A partial moon shed a weak light over the scene and revealed the foreman entering the other tent with the axe in his hand. A short silence was followed by a heavy splat, a tremendous grunt, and some frenzied shouts. The supply tent heaved violently, went down, and split open as the bear hurtled out and through the woods toward the creek. When order had been restored it transpired that the foreman had stolen up to the bear, which had its back to him, and had struck it across the rump as hard as he could with the flat of the axe. The element of surprise apparently was all in his favor because the startled bear charged directly away from him into the far end of the tent. Although in this instance no injuries were suffered, it has always seemed that this was an extremely foolhardy thing to do. Although one of the most laughable happenings I have ever seen, it also had all the elements of a possible tragedy.

Grizzly bear
Ursus horribilis (Latin: a bear ... horrible)

RANGE: Alaska, western Canada, and in the United States confined to the high mountains of the Continental Divide as far south as northern New Mexico.

HABITAT: Except in National Park areas, grizzlies are seldom seen, since they frequent only the most isolated places in the mountains; Transition Life Zone and higher.

DESCRIPTION: The largest carnivore in the Southwest. Easily distinguished from the black bear by the prominent hump on the shoulders. Total length 6 to 7 feet. Tail so short as to be unnoticeable. Height at shoulders 3 to 3½ feet. Weight 325 to 850 pounds. Color of the southwestern grizzlies is variable, ranging from yellowish brown to nearly black, but has a characteristic grizzled effect caused by the white-tipped hairs scattered through the fur. This is especially noticeable along the back. The grizzly, though massively built, gives an impression of leanness. The shoulders are higher than the posterior, giving the animal a streamlined appearance. The head is large and round with a square, uptilted muzzle. The legs are extremely powerful, the feet large and with formidable claws, those of the front feet being up to 4 inches long. The young will number from one to three, with two being most common. Grizzlies breed every 2 or 3 years.

Probably no mammal in the United States is more certain soon to become extinct than these great bears. Many factors contribute toward this end, chief among them being the low reproduction rate and the rapid decrease of its range because of an increase in stock raising and agriculture. Ousted from its former haunts, the species is now found chiefly in only the few areas where it is rigidly protected. It seems extremely unlikely that it can long survive this reduction of its once unlimited range. This is the culmination of a program of destruction wrought on the grizzly since penetration of the white man into the West. It but follows the disappearance of other, less well known bears which lived in the Southwest at that time.

When the Mountain Men came into the West in the period from 1800 to 1850 they found a huge, light-colored bear roaming the foothills of the desert country. For want of a better name they called it the "gray bear." From the accounts of that time it is now assumed that it was a grizzly; at any rate, it was said to have been extremely ferocious, a trait which led to its downfall. In the space of about 70 years this animal was discovered, hunted and exterminated, all without a specimen of any kind being preserved. Today not a trace of this big predator remains. Its fate illustrates the usual result of contact between a dangerous, highly specialized animal and man. The question which arises is, should any group of men ever be allowed such control over a wilderness that they are able to exterminate the fauna and flora to the detriment of succeeding generations? The answer seems obvious if we consider that "we but hold these things in trust."

grizzly bear

Many species of the grizzly are recognized by taxonomists, but few are alive today. In the United States only New Mexico, Colorado, Utah, Wyoming, Montana, and Idaho still have some of these big animals. In some other western States they have but recently become extinct. California is thought to have lost its last grizzly in 1925. The few survivors are probably all of the species *horribilis*. Since grizzly country is also black bear country, the layman may become confused in identifying the two species. A few important differences make identification easy.

The first and most conspicuous field mark is the prominent shoulder hump of the grizzly. The male black bear will sometimes with age develop a shoulder hump, but it cannot compare with that of the grizzly. Second, the grizzly has what has been described as a "dish" face; that is, a concavity in the general shape of the front of the face, whereas the black bear develops a definite "Roman" nose. Third, the claws of the grizzly are twice as long as those of the black bear; this is most noticeable in the tracks. If one is close enough to see this characteristic in the field, he probably is too close for safety! Lastly, the attitude of the two species toward each other when they meet on common ground is characteristic. As a rule, the approach of a grizzly to a garbage dump is enough to put all black bears to flight. There is no intermingling of the two species; the grizzly is the master and the black bear will not challenge his authority.

In most of its habits the grizzly resembles the black bear. It is omnivorous to the same degree, but somewhat more predatory. It also goes into hibernation for the winter, and the cubs are born during this inactive period. They receive the same rigorous training as that accorded their black cousins, and like them, are able to climb into the trees and out of harm's way. As they grow older, this ability leaves them with the growing of the long claws, and adult grizzlies are supposed to be unable to climb. In one respect the grizzly differs from not only the black bear but from most other native mammals. It has never learned to fear man to the same degree that other creatures have.

Whether the grizzly's belligerent attitude stems from fear or contempt is a moot question. The important point to remember is that a grizzly should be avoided at all times. Injuries suffered by humans in their contacts with black bears are usually accidental rather than the result of deliberate assault by the animal. Grizzlies have been known to charge without other provocation than trespass on what they consider their territory. Surely the public can afford to humor this irascible giant. A little consideration for its irritable nature is not too great a price to pay for its continued existence in our rapidly dwindling numbers of large carnivores.

Vagrant shrew
Sorex vagrans (Latin: a shrew ... wandering)

RANGE: Confined to mountains of western United States and Canada, and northern and southern Mexico.

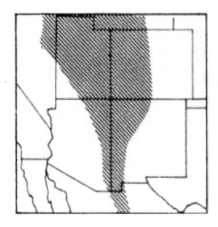

HABITAT: Moist places in forests of the Transition Life Zones and higher.

DESCRIPTION: A tiny creature with a long nose. Total length 4 to 5 inches. Tail 1½ to 2 inches. Color reddish brown to black above with sides drab and lightening to gray below. Tail indistinctly bicolor except for the last half which is dark all the way around. Head round and narrowing to a long, pointed, somewhat flexible nose. Long whiskers are found along the sides of the upper jaw. Eyes and ears so small as to be difficult to see. Little is known of breeding habits of the shrews. The vagrant shrew is said to breed at any time of year and to have from 5 to 11 young in a litter.

Shrews are the smallest American mammals. Their size and secretive habits combine to make them among the least known of native animals. They are classed as insectivores, although they eat other small mammals as well as insects. They may be distinguished from mice by their bicuspid incisors and modified canine teeth. Another difference is that shrews have five toes, in contrast to the four-toed feet of mice.

vagrant shrew

As far as is known at present, certain species of shrews are the only poisonous mammals. The big short-tailed shrew (eastern United States) has a toxic substance in its saliva which helps subdue some of the animals it captures. It is thought that some western species also have this peculiarity. Though shrews are among the tiniest animals known, they are not unduly persecuted by larger predators. This is thought to be partly because of certain glands on the shrew's body which give it an offensive odor.

An outstanding characteristic of shrews is their need for a constant supply of food. Because all small animals lose heat quickly, they must eat almost constantly to replace this loss. Some species will eat their own weight in food as often as every 3 hours. An outstanding exception is the water shrew, which can do without food for as long as 2 days without starving to death. Since most shrews live in or near the water, they find ample food in the insects, spiders, minnows, and small mammals which live in moist locations. The group is as ferocious as it is voracious. Most shrews do not hesitate to attack animals outweighing them several times. It has been said that if shrews were as big as squirrels they probably would even attack man.

In the mountains of Utah, Colorado, and northern New Mexico the northern water shrew (*Sorex palustris*) may be encountered. It is somewhat larger than the vagrant shrew and will not be seen away from water. Gray below and black above, it is wonderfully camouflaged, whether in water or on land. It, like other shrews, has long whiskers known as vibrissae. Land shrews use these whiskers as tactile organs to help them follow the dark maze of their runways. Water shrews are thought to use them as sense organs in place of eyes to pursue the minnows, tadpoles, and water bugs they eat. Actually, the water shrew resembles a large water bug as it darts about below the surface surrounded by the silvery bubbles of air imprisoned in its fine fur.

Bats
Order *Chiroptera* (Latin: chir, hand, and optera, wing)

The special treatment accorded bats in this book is not given them by choice. It results from an inability to so clearly describe any one or two species chosen that the layman might be able to distinguish these from their numerous and equally interesting relatives. When one considers that numerically bats are thought to compare favorably with birds, that there are a great number of species divided into many genera, and that the four-State area with which we are concerned is invaded, so to speak, by eastern, northern, western and Mexican species besides having several of its own, it soon becomes apparent that this group can be described here only in the most general way. If some of the popular superstitions about bats are contradicted here, it is to be hoped the reader will find the facts no less interesting.

The adaptation for which bats are best known is their ability to fly. This specialized talent is shared by no other type of mammal. It is made possible by considerable modification of several structures of the body, that of the forelimbs being the most extreme. The bones of both the upper and lower forelegs are considerably lengthened, but cannot compare with the extreme elongation of the digits. The clawlike protuberance from the front of the wing corresponds to the thumb. The wing membrane stretched across the "fingers" is attached to the side of the body and to the hindlegs as far as the ankle. Most bats have another wing membrane, called the interfemoral membrane, which joins both hind legs, and in many species it also embraces the tail. The wing membranes look and feel somewhat like thin leather. Running through them is an intricate system of blood vessels. These not only supply nourishment to the membrane but also act as a radiator in cooling the blood stream during the strenuous physical labor involved in flight. The principles of flight are similar to those used by birds; that is, the wings are partially folded on the upstroke and fully extended during the down beat. This maneuver produces a rustling sound that is clearly audible in the quiet of a cave. In fact, if thousands of bats are disturbed at the same time it becomes a low roar.

The fact that bats are nocturnal, and at the same time lead an aerial life which necessitates flying through labyrinths plunged in total darkness, has been the cause of much research as to the means by which they can do this. It is now definitely known that they depend on a sonar system where, by emitting shrill cries, they are guided by the echoes rebounding from nearby objects. These "squeaks" range within a frequency of from 25,000 to 75,000 vibrations per second, which is too high for the human ear to register. The sounds are uttered at rates from about 10 per second when the bat is at rest to as many as 60 per second when it is in flight and surrounded by the many obstacles to be found in a cave. Fantastic as this performance seems, it is matched by a theory that tiny muscles close the bat's ears to each squeak and open them again to hear only the echo.

The response of their vocal and hearing structure to this specialized use is truly amazing. There are no more unique faces in the mammal kingdom than those of the bats. Most bats have enormous ears with ridged and channeled interiors that probably have much to do with amplifying faint sounds. Set in front of the ear is a narrow, upright protuberance known as the tragus. Farther down the face, in the region of the nose, are other strangely shaped skin structures including the "nose leaf." As yet the functions of these appendages are not entirely known, but it is suspected that at least part of their purpose is to beam the squeaks along a definite line and thus help orient the bat with its surroundings. With such an efficient system to guide it, the bat has small need for eyes. The expression "blind as a bat" is misleading, however, because most bats, in spite of their relatively small eyes, can see rather well.

Since most southwestern bats are insectivorous, with the exception of a very few species along the Mexican border which are considered fruit eaters, the question arises as to how they exist during the winter months when insects are not to be found. There are two common methods by which animals avoid such a lean period: by migration and by hibernation. Bats employ both. Some species are thought to fly as far south as Central America. Others group together in caves and hang in a deep torpor all winter. In this state of inactivity their body temperatures may fall to within one degree of their surroundings, and their rate of metabolism sometimes falls to one-eighteenth of that during active periods. As a rule, bats prefer a cool place for hibernation, because the cooler the temperature the slower the rate of metabolism. Body temperatures as low as 33.5° F. have been recorded in hibernating bats. The temperature must not fall below freezing, or the animals will perish. During this period of inactivity bats have been known to lose up to one-third of their weight.

Because of their secretive habits and nocturnal periods of activity, bats have few enemies other than man that are capable of making any serious inroads on their numbers. Consequently the birth rate is quite low in most species. Many have no more than one young each year; and the red bat, which bears up to four young, seems to be the most prolific in the United States. There is great variety in the methods by which different species care for the young. Some mothers leave the babies hanging to the roof of the cave while they go on their nightly search for food; others carry the young clinging tightly to their fur. The young mature quickly. They are usually able to fly within a month after they are born.

> Despite much recent scientific study, bats are still among our least known creatures. Their insectivorous diet surely makes them of great importance to man. Beyond this, their immense numbers indicate that ecologically they must have tremendous influence on any area in which they live.

CPSIA information can be obtained
at www.ICGtesting.com
Printed in the USA
LVHW040555081222
734780LV00030B/991